Advances in Industrial Control

Other titles published in this Series:

Digital Controller Implementation and Fragility
Robert S.H. Istepanian and James F. Whidborne (Eds.)

Optimisation of Industrial Processes at Supervisory Level
Doris Sáez, Aldo Cipriano and Andrzej W. Ordys

Robust Control of Diesel Ship Propulsion
Nikolaos Xiros

Hydraulic Servo-systems
Mohieddine Jelali and Andreas Kroll

Strategies for Feedback Linearisation
Freddy Garces, Victor M. Becerra, Chandrasekhar Kambhampati and Kevin Warwick

Robust Autonomous Guidance
Alberto Isidori, Lorenzo Marconi and Andrea Serrani

Dynamic Modelling of Gas Turbines
Gennady G. Kulikov and Haydn A. Thompson (Eds.)

Control of Fuel Cell Power Systems
Jay T. Pukrushpan, Anna G. Stefanopoulou and Huei Peng

Fuzzy Logic, Identification and Predictive Control
Jairo Espinosa, Joos Vandewalle and Vincent Wertz

Optimal Real-time Control of Sewer Networks
Magdalene Marinaki and Markos Papageorgiou

Process Modelling for Control
Benoît Codrons

Computational Intelligence in Time Series Forecasting
Ajoy K. Palit and Dobrivoje Popovic

Modelling and Control of mini-Flying Machines
Pedro Castillo, Rogelio Lozano and Alejandro Dzul

Rudder and Fin Ship Roll Stabilization
Tristan Perez

Hard Disk Drive Servo Systems (2nd Ed.)
Ben M. Chen, Tong H. Lee, Kemao Peng and Venkatakrishnan Venkataramanan

Measurement, Control, and Communication Using IEEE 1588
John Eidson

Piezoelectric Transducers for Vibration Control and Damping
S.O. Reza Moheimani and Andrew J. Fleming

Manufacturing Systems Control Design
Stjepan Bogdan, Frank L. Lewis, Zdenko Kovačić and José Mireles Jr.

Windup in Control
Peter Hippe

Nonlinear H_2/H_∞ Constrained Feedback Control
Murad Abu-Khalaf, Jie Huang and Frank L. Lewis

Practical Grey-box Process Identification
Torsten Bohlin

Modern Supervisory and Optimal Control
Sandor Markon, Hajime Kita, Hiroshi Kise and Thomas Bartz-Beielstein

Wind Turbine Control Systems
Fernando D. Bianchi, Hernán De Battista and Ricardo J. Mantz

Advanced Fuzzy Logic Technologies in Industrial Applications
Ying Bai, Hanqi Zhuang and Dali Wang (Eds.)

Practical PID Control
Antonio Visioli

Advanced Control of Industrial Processes
Piotr Tatjewski
Publication due October 2006

Adaptive Voltage Control in Power Systems
Giuseppe Fusco and Mario Russo
Publication due October 2006

Luigi Fortuna, Salvatore Graziani,
Alessandro Rizzo and Maria G. Xibilia

Soft Sensors for Monitoring and Control of Industrial Processes

With 179 Figures

Springer

Luigi Fortuna, Prof., Eng.
Università degli Studi di Catania
Dipartimento di Ingegneria Elettrica
Elettronica e dei Sistemi
95125 Catania
Italy

Salvatore Graziani, Prof., Eng., Ph.D.
Università degli Studi di Catania
Dipartimento di Ingegneria Elettrica
Elettronica e dei Sistemi
95125 Catania
Italy

Alessandro Rizzo, Dr., Eng., Ph.D.
Politecnico di Bari
Dipartimento di Elettrotecnica
ed Elettronica
70125 Bari
Italy

Maria G. Xibilia, Dr., Eng., Ph.D.
Università degli Studi di Messina,
Facoltà di Ingegneria
Dipartimento di Matematica
98166 Messina
Italy

British Library Cataloguing in Publication Data
Soft sensors for monitoring and control of industrial
processes. - (Advances in industrial control)
1.Detectors - Design 2.Manufacturing processes -
Mathematical models 3.Process control 4.Electronic
instruments 5.Engineering instruments
I.Fortuna, L. (Luigi), 1953-
681.2
ISBN-13: 9781846284793
ISBN-10: 1846284791

Library of Congress Control Number: 2006932285

Advances in Industrial Control series ISSN 1430-9491
ISBN-10: 1-84628-479-1 e-ISBN 1-84628-480-5
ISBN-13: 978-1-84628-479-3

Printed on acid-free paper

9 8 7 6 5 4 3 2 1

Springer Science+Business Media
springer.com

Advances in Industrial Control

Professor Emeritus O.P. Malik
Department of Electrical and Computer Engineering
University of Calgary
2500, University Drive, NW
Calgary
Alberta
T2N 1N4
Canada

Professor K.-F. Man
Electronic Engineering Department
City University of Hong Kong
Tat Chee Avenue
Kowloon
Hong Kong

Professor G. Olsson
Department of Industrial Electrical Engineering and Automation
Lund Institute of Technology
Box 118
S-221 00 Lund
Sweden

Professor A. Ray
Pennsylvania State University
Department of Mechanical Engineering
0329 Reber Building
University Park
PA 16802
USA

Professor D.E. Seborg
Chemical Engineering
3335 Engineering II
University of California Santa Barbara
Santa Barbara
CA 93106
USA

Doctor K.K. Tan
Department of Electrical Engineering
National University of Singapore
4 Engineering Drive 3
Singapore 117576

Professor Ikuo Yamamoto
Kyushu University Graduate School
Marine Technology Research and Development Program
MARITEC, Headquarters, JAMSTEC
2-15 Natsushima Yokosuka
Kanagawa 237-0061
Japan

Series Editors' Foreword

The series *Advances in Industrial Control* aims to report and encourage technology transfer in control engineering. The rapid development of control technology has an impact on all areas of the control discipline. New theory, new controllers, actuators, sensors, new industrial processes, computer methods, new applications, new philosophies..., new challenges. Much of this development work resides in industrial reports, feasibility study papers and the reports of advanced collaborative projects. The series offers an opportunity for researchers to present an extended exposition of such new work in all aspects of industrial control for wider and rapid dissemination.

The rapid invasion of industrial and process control applications by low-cost computer hardware, graphical-user-interface technology and high-level software packages has led to the emergence of the virtual instrumentation paradigm. In fact, some manufacturers quickly recognised the potential of these different aspects for exploitation in producing virtual instrumentation packages and modules as exemplified by the LabVIEW™ product from National Instruments.

As this monograph makes clear, virtual instrumentation is a computer-based platform of hardware and software facilities that can be used to create customised instruments for a very wide range of measurement tasks. These facilities involve: a user interface to enable the flexible construction, operation and visualisation of the measurement task; computational software to allow advanced processing of the measurement data; and software to integrate hardware units and sensors into the virtual instrument and to orchestrate their operation.

By way of comparison, Professor Fortuna and his colleagues consider "soft sensors" to be a far narrower concept within the topic of virtual instrumentation, stating that "soft sensors focus on the process of estimation of any system variable or product quality by using mathematical models, substituting some physical sensors and using data acquired from some other available ones." Thus, the methods in this *Advances in Industrial Control* monograph have very strong links to the procedures of industrial-process-model identification and validation.

The monograph opens with three chapters that establish the background to soft sensors; this presentation culminates in Chapter 3 where the complete design process for these sensors is described. Chapters 4, 5 and 6 are then sharply focussed on the key steps in soft sensor design: data selection; model structure selection and model validation, respectively. Extensions to the basic steps of soft-sensor design, namely soft-sensor performance enhancement and the modifications needed to facilitate different industrial process applications follow in Chapter 7 and 8, respectively. Widening the applications range and role of soft sensors to fault detection and sensor validation configurations is dealt with in Chapter 9.

A great strength of *Soft Sensors for Monitoring and Control of Industrial Processes* is the use, throughout the text, of a set of industrial case studies to demonstrate the successes and drawbacks of the different methods used to create soft-sensor models. A number of different methods may be used in each separate step of the soft-sensor design process and the industrial case studies are often used to provide explicit comparisons of the performance of these methods. The industrial control and process engineer will find these comparison exercises invaluable illustrations of the sort of results that might be found in industrial applications.

The monograph also highlights the importance of using knowledge from industrial experts and from the existing industrial process literature. This is an important aspect of industrial control that is not very widely acknowledged or taught in control courses. Most industrial processes have already generated a significant experimental knowledge base and the control engineer should develop ways of tapping into this valuable resource when designing industrial control schemes.

This is a monograph that is full of valuable information about the veracity of different methods and many other little informative asides. For example, in Chapter 9, there is a paragraph or two on trends in industrial applications. This small section seeks to determine whether and how nonlinear models are used in industrial applications. It presents some preliminary data and argument that "the number of nonlinear process applications studied through nonlinear models has been clearly increasing over the years, while nonlinear process applications with linearised models have been decreasing." A very interesting finding that deserves further in-depth investigation and explanation.

The industrial flavour of this monograph on soft sensors makes it an apposite volume for the *Advances in Industrial Control* series. It will be appreciated by the industrial control engineer for its practical insights and by the academic control researcher for its case-study applications and performance comparisons of the various theoretical procedures.

M.J. Grimble and M.A. Johnson
Glasgow, Scotland, U.K.

Preface

This book is about the design procedure of soft sensors and their applications for solving a number of problems in industrial environments.

Industrial plants are being increasingly required to improve their production efficiency while respecting government laws that enforce tight limits on product specifications and on pollutant emissions, thus leading to ever more efficient measurement and control policies. In this context, the importance of monitoring a large set of process variables using adequate measuring devices is clear. However, a key obstacle to the implementation of large-scale plant monitoring and control policies is the high cost of on-line measurement devices.

Mathematical models of processes, designed on the basis of experimental data, via system identification procedures, can greatly help, both to reduce the need for measuring devices and to develop tight control policies. Mathematical models, designed with the objectives mentioned above, are known either as virtual sensors, soft sensors, or inferential models.

In the present book, design procedures for virtual sensors based on data-driven approaches are described from a theoretical point of view, and relevant case studies referring to real industrial applications, are described. The purpose of the book is to provide undergraduate and graduate students, researchers, and process technologists from industry, a monograph with basic information on the topic, suggesting step-by-step solutions to problems arising during the design phase. A set of industrial applications of soft sensors implemented in the real plants they were designed for, is introduced to highlight their potential.

Theoretical issues regarding soft sensor design are illustrated in the framework of specific industrial applications. This is one of the valuable aspects of the book; in fact, it allows the reader to observe the results of applying different strategies in practical cases. Also, the strategies adopted can be adapted to cope with a large number of real industrial problems.

The book is self-contained and is structured in order to guide the interested reader, even those not closely involved in inferential model design, in the development of their own soft sensors.

Moreover, a structured bibliography reporting the state of the art of the research into, and the applications of, soft sensors is given.

All the case studies reported in the book are the result of collaboration between the authors and a number of industrial partners. Some of the soft sensors developed are implemented on-line at industrial plants.

The book is structured in chapters that reflect the typical steps the designer should follow when developing his own applications. The reader can refer to the following scheme as a guide with which to search the book for solutions to particular aspects of a typical soft sensor design. Also, soft sensor design procedure is not straightforward and the designer sometimes needs to reconsider part of the design procedure. For this reason, in the scheme, a path represented by grey lines overlaps the book structure to represent possible soft sensor design evolution.

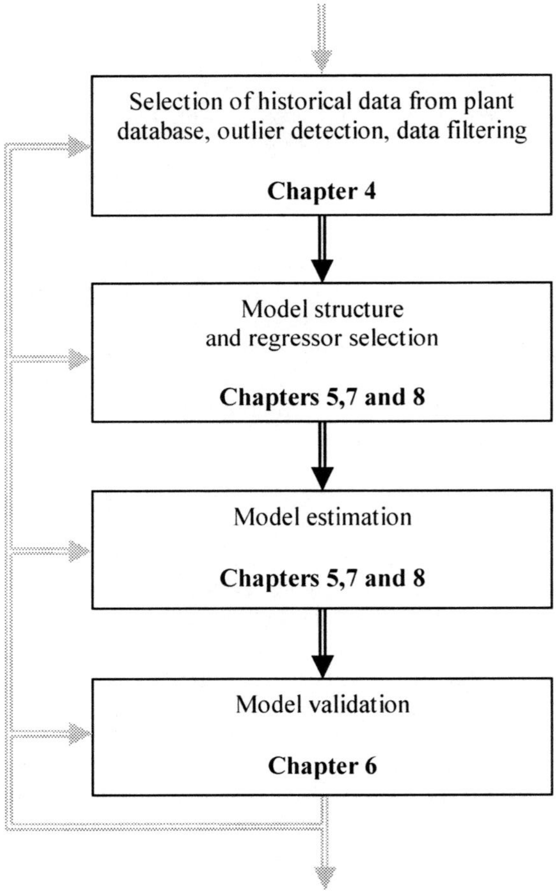

The state of the art on research into, and industrial applications of, soft sensors is reported in Chapter 1. Chapters 2 and 3 give some definitions and a short description of theoretical issues concerning soft sensor design procedures. Chapter 9 deals with the related topic of model-based fault detection and sensor validation, giving both the state of the art and two applications of sensor validation. Technical details of plants used as case studies are reported in the Appendix A.

As a complement to the bibliography section, where works cited in the book are listed, a structured bibliography is provided, in Appendix B, with the aim of guiding the reader in his or her search for contributions on specific aspects of soft sensor design.

Readers wishing to apply the techniques for soft sensor design described in the book will find data taken from real industrial applications in the book web site: www.springer.com/1-84628-479-1.

Catania, March 2006

Luigi Fortuna
Salvatore Graziani
Alessandro Rizzo
M. Gabriella Xibilia

Acknowledgments

We are most grateful to all those from industry and research laboratories, not forgetting our colleagues, who have been working with us for many years of research in this field. In particular, our special thanks go to Bruno Andò, Giuliano Buceti, Paolo Debartolo, Giovanni Di Battista, Vito Marchese, Peppe Mazzitelli, and Mario Sinatra.

Thanks are also due to Tonino Di Bella and Pietro Giannone, who helped with graphics and simulations.

Finally, we are indebted to those who helped us in a number of different ways: Doretta and Lina, Giovanna and Gaetano, Michele, Pippo and Meluccia, Francesca, Mario, Sara Eva, and Arturo.

Contents

1

Soft Sensors in Industrial Applications

1.1 Introduction

Soft sensors are a valuable tool in many different industrial fields of application, including refineries, chemical plants, cement kilns, power plants, pulp and paper industry, food processing, nuclear plants, urban and industrial pollution monitoring, just to give a few examples. They are used to solve a number of different problems such as measuring system back-up, what-if analysis, real-time prediction for plant control, sensor validation and fault diagnosis strategies.

This book deals with some key points of the soft sensors design procedure, starting from the necessary critical analysis of rough process data, to their performance analysis, and to topics related to on-line implementation.

All the aspects of soft sensor design are dealt with both from a theoretical point of view, introducing a number of possible approaches, and with numerical examples taken from real industrial applications, which are used to illustrate the behavior of each approach.

Industries are day by day faced with the choice of suitable production policies that are the result of a number of compromises among different constraints. Final product prices and quality are of course two relevant and competing factors which can determine the market success of an industry. Strictly related to such aspects are topics like power and raw materials consumption, especially because of the ever growing price of crude oil. Moreover, the observance of safety rules (according to several studies, inadequate management of abnormal situations represents a relevant cause of loss in industry) and environmental pollution issues contribute to increase the complexity of the outlined scenario.

In recent decades, people and politicians have focused their attention on these topics, and regulations have been promoted by governments. Companies are required to respect laws that enforce more and more strict limits on product specifications and pollutant emissions of industrial plants.

A relevant example is the Kyoto treaty, which is a legal agreement under which industrialized countries agreed to reduce their collective emissions of greenhouse gases by 5.2% compared to the year 1990. The goal of the treaty is to lower overall emissions of six greenhouse gases – carbon dioxide (CO_2), methane (CH_4), nitrous oxide (N_2O), sulfur hexafluoride (SF_6), hydrofluorocarbons (HCFs), and perfluorocarbons (PFCs) – calculated as an average over the five-year period 2008–12. The treaty came into force on February 16, 2005 following ratification by Russia on November 18, 2004. As of September 2005, a total of 156 countries have ratified the agreement (representing over 61% of global emissions). Although notable exceptions include the United States and Australia, the agreement clearly shows that environmental issues are recognized as global problems.

The constraints mentioned above represent a continuous challenge for process engineers, politicians and operators; adequate solutions require a deep, quantitative knowledge of the process and of relevant process parameters. The importance of monitoring a large set of process variables by installing and using adequate measuring systems (generally in the form of distributed monitoring networks) is therefore clear.

Unfortunately measuring devices are generally required to work in a hostile environment that, on the one hand, requires instrumentation to meet very restrictive design standards, while on the other hand a maintenance protocol has to be scheduled. In any case, the occurrence of unexpected faults cannot be totally avoided. Nevertheless, some measuring tools can introduce a significant delay in the application that can reduce the efficiency of control policies. To install and maintain a measuring network devoted to monitoring a large plant is never cheap and the required budget can significantly affect the total running costs of the plant, which are generally biased to reduce the total number of monitored variables and/or the frequency of observations, though in many industrial situations infrequent sampling (lack of on-line sensors) of some process variables can present potential operability problems. A typical case is when variables relevant to product quality are determined by off-line sample analyses in the laboratory, thus introducing discontinuity and significant delays (Warne *et al.*, 2004).

Cases can be mentioned where it is impossible to install an on-line measuring device because of limitations of measuring technologies. Also in such cases the variables that are key indicators of process performance are determined by off-line laboratory analyses.

Mathematical models of processes designed to estimate relevant process variables can help to reduce the need for measuring devices, improve system reliability and develop tight control policies.

Plant models devoted to the estimation of plant variables are known either as inferential models, virtual sensors, or soft sensors.

Soft Sensors offer a number of attractive properties:

- they represent a low-cost alternative to expensive hardware devices, allowing the realization of more comprehensive monitoring networks;
- they can work in parallel with hardware sensors, giving useful information for fault detection tasks, thus allowing the realization of more reliable processes;
- they can easily be implemented on existing hardware (*e.g.* microcontrollers) and retuned when system parameters change;

- they allow real-time estimation of data, overcoming the time delays introduced by slow hardware sensors (*e.g.* gas chromatographs), thus improving the performance of the control strategies.

There are three main approaches to building soft sensors: mechanistic modeling (physical modeling), multivariate statistics, and artificial intelligence modeling such as neural networks, fuzzy logic and hybrid methods. This classification approach is not intended to be very rigid, and methodologies typical of one of them are often improved by techniques typical of others.

Suitable empirical models, or data-driven models, producing reliable real-time estimates of process variables on the basis of their correlation with other relevant system variables can be useful tools in industrial applications, due to the complexity of the plant dynamics, which can prevent the first principles approach from being used.

The accumulated historical record generally collected by industries in fact represents a useful source of information, which can enable relevant features to be identified (Albazzaz and Wang, 2006).

However, the potential information regarding factors affecting plant operation might be obscured by the sheer volume of data collected (Flynn, Ritchie and Cregan, 2005). Moreover, the process of data mining can be difficult because of high dimensionality, noise and low accuracy, redundant and incorrect values, non-uniformity in sampling and recording policies.

The importance of data collection policy and critical analysis of available data can never be emphasized enough. Data collection is a fundamental issue because a model cannot be better than the data used for its estimation: poor results are generally obtained if collected data are passed on without any action, such as selection, filtering, *etc.*, to some modeling procedure. The model designer might select data that represent the whole system dynamic when this is possible by running suitable experiments on the plant. Effects of disturbances should also be filtered out.

Moreover, careful investigation of available data is required in order to detect either missing data or outliers, due to faults of measuring or transmission devices or to unusual disturbance, which can have unwanted effects on model quality. In fact, any help from plant experts should be considered a precious support to any numerical data processing approach.

Collected data can be processed in different ways to design the soft sensor. A number of choices are necessary in order to select both the model class (*e.g.* linear or nonlinear, static or dynamic, and so on) and the identification approach most suitable to the problem under investigation.

The last step in soft sensor design, *i.e.* the problem of model validation, can be approached using a number of different strategies.

All the aspects mentioned will be described in detail in the following chapters through a number of industrial case studies.

1.2 State of the Art

The literature on soft sensors in industrial applications, concerning both theoretical and practical aspects, consists of a number of very specialized journals, international conferences, and workshops. Nevertheless some theoretical aspects related to modeling, signal processing, and identification theory can be found in books and conferences devoted to system theory, automatic control, instrumentation and measurement, and artificial intelligence.

It is easy to understand that any attempt to give an exhaustive description of such a huge literature would necessarily be unsuccessful. Therefore, we will proceed in what follows, to describe the state of the art, referring to relevant contributions and trying to give an order to the referenced material, by using some classification criteria. In the case of reported applications, we will refer mostly to recent literature.

The present survey is not intended to be exhaustive, and obviously classification schemes different from the proposed one are possible. In addition, class boundaries should be considered as somewhat fuzzy and overlapping: it is not always possible to focus on one single aspect without addressing correlated ones.

A first classification of the relevant literature will be between theoretical and applicative contributions. For the former class, further classification will follow the typical steps of soft sensor design and can be summarized as follows:

- data collection and filtering;
- variables and model structure selection;
- model identification;
- model validation.

Some books are available that address some of the steps mentioned. The book by Ljung (1999) is considered a milestone in the field of identification theory. A valuable source of theoretical information on linear multiple input–multiple output (MIMO) system identification can be found in Guidorzi (2003).

Though most industrial processes should be better identified by nonlinear models, there are very few books devoted explicitly to this topic. Among these, that by Nørgaard *et al.* (2000) deals with nonlinear models, implemented using neural networks. The known approximation property of some neural network structures is exploited by the authors to obtain the nonlinear generalization of linear model structures. In particular, relevant topics like design of the input signals for experiments, data collection and pre-processing, lag selection, parameter identification (in the form of neural network training strategies), regularization, model structure adaptation (neural network pruning) and model validation are dealt with.

Also of interest is the book by Omidvar and Elliott (1997), where one chapter is devoted to identification of nonlinear dynamic systems using neural networks and another deals with practical issues regarding the use of neural networks for intelligent sensors and control.

In recent years a number of books have been published dealing with soft computing and artificial intelligence techniques. Some aspects of these fields form the basis of the approaches reported in this book. Readers who have no in-depth knowledge of this topic can refer to Haykin (1999), Fortuna *et al.* (2001), or Gupta and Sinha (2000).

These books deal with theoretical and practical aspects of soft sensors, while little attention is given to real case studies. In contrast, in the present book we focus attention mainly on real industrial applications, without dealing in depth with theoretical issues. Readers interested in theoretical aspects can refer to the reported bibliography.

1.2.1 Data Collection and Filtering

Large industries are generally required to collect and store data on sensitive process parameters, and the same holds for large cities as regards pollutant levels. This paves the road to the subsequent use of data for model identification. Unfortunately data collection strategies sometimes do not fit the requirements of identification techniques (*e.g.* problems can arise with sampling time, missing data, outliers, working conditions, accuracy and so on).

The strategy adopted for data collection, and the critical analysis of available data are fundamental issues in system identification. The very first issue to be addressed concerns with the sampling frequency, which depends on the system dynamics. Plenty of books deal with the process of data sampling for continuous time systems. A good example of a book dedicated to such a topic is that by Oppenheim and Schafer (1989), where sampling theory is addressed together with correlated topics such as anti-alias filtering, signal reconstruction and so forth.

An in-depth description of the negative impact of data compression policies, often adopted in industrial plants to enable storage cost reduction, can be found in Thornhill *et al.* (2004), while the effect of the presence of missing data in the historical plant database, deriving from failure in sensors, is dealt with in Lopes and Menezes (2005), where projection to latent structures (PLS) models are used to develop a soft sensor for industrial petrochemical crude distillation columns. Principal component analysis (PCA) and PLS methods in the case of missing data are also dealt with in Nelson, Taylor and MacGregor (1996).

Another relevant topic regarding collected data quality is the presence of outliers, resulting from hardware failure, incorrect readings from instrumentation, transmission problems, ‘strange’ process working conditions, and so on. Different techniques for outlier detection are reviewed in Warne *et al.* (2004), Englund and Verikas (2005), Lin *et al.* (2005), Pearson (2002), and Chiang, Pell and Seasholtz (2003). In particular, in Englund and Verikas (2005) a survey of methods for outlier detection is reported along with a new strategy which aggregates different approaches. The proposed approach is applied to the design of a soft sensor for an offset lithographic printing process.

After outliers have been successfully detected, data may still be inadequate for soft sensor design, and operations, generally known as pre-filtering, are required. A general treatment of the role of pre-filtering in model identification can be found in Ljung (1999) and Guidorzi (2003). The role of pre-filtering in nonlinear system identification is analyzed in Spinelli, Piroddi and Lovera (2005), where a frequency domain interpretation is provided based on the use of the Volterra series representation.

1.2.2 Variables and Model Structure Selection

Different strategies have been proposed in the literature to model real systems depending on the level of *a priori* knowledge of the process. Models can be obtained either on the basis of first principles analysis (also known as mechanistic models) or by using gray- or black-box identification approaches.

In the case of processes involved in industrial plants, due to the complexity of the phenomena involved, mechanistic modeling can be very time consuming and significant parameters are generally unknown. However, the great amount of historical data, usually acquired for monitoring purposes, suggests the use of nonlinear gray- or black-box process model identification.

Even if it is difficult to give a theoretical treatment of the gray-box approach (it essentially depends on both the type of process under investigation and the level of available physical insight), contributions do exist on practical applications. The gray-box approach can lead to very accurate models because it exploits any available source of information to refine the model.

Two recent contributions describing industrial applications are those of Zahedi *et al.* (2005), and Van Deventer, Kam and Van der Walt (2004). In the former, a hybrid model of the differential catalytic hydrogenation reactor of carbon dioxide to methanol is proposed. The model consists of two parts: a mechanistic model and a neural one. The mechanistic model calculates the effluent temperature of the reactor by taking outlet mole fractions for a neural model. The authors show that the hybrid model outperforms both a first principles model and a neural network model using the available experimental data. A set of other interesting applications of the gray-box approach can be found in the reference list of the paper.

The paper by Van Deventer, Kam and Van der Walt (2004) is an example of an effort to include prior knowledge of a process into neural models in such a way that the interactions between the process variables are represented by the network's connections by means of regression networks. A regression network is a framework by which a model structure can be represented using a number of feedforward interconnected nodes, each characterized by its own transfer function. In particular, the dynamic modeling of continuous flow reactors using the carbon-in-leach process for gold recovery is proposed as a case study. Black-box regression techniques are compared to the regression network and the latter is shown to give better performances.

The present book focuses mainly on the black-box approach because it can give satisfactory results in complex industrial modeling applications, with reasonable computational and time efforts. In what follows, we will report significant examples of different identification techniques devoted to black-box modeling.

The aspects of variable and model structure selection are of key importance and therefore they are widely investigated in the literature, even if it is hard to find a general solution that clearly outperforms others. This outlines a fundamental aspect of black-box modeling: any technologist knowledge, regarding the input variable choice, the system order, the operating range, time delay, degree of nonlinearity, sampling times, *etc.*, represents a valuable source of information that should be taken into account by the model designer. This is very true when nonlinear systems are considered. Though most of the literature deals with theoretical results for linear model identification, we will focus our attention on literature about nonlinear applications that, in our opinion, are the closest to reality.

The very first question a model designer is faced with regards the choice of independent variables that influence the model output. In Warne *et al.* (2004) the authors, among other topics, reviewed a number of techniques that can be used for linear and nonlinear modeling problems. The first and most intuitive approach to the problem of variable selection discussed in this paper is the graphical inspection of dispersion plots aimed at discovering any structure in the graphs obtained. Moreover, more quantitative criteria such as the coefficient of correlation and Mallows' statistics are reviewed.

In Rallo *et al.* (2002), Kohonen maps are used to solve the same problem. Kohonen maps belong to the class of self-organizing maps. In the application mentioned self-organizing maps are used to project subsets of input variables along with the output variables onto network output space. A dissimilarity method is used to determine the relevance of each combination. The proposed strategy is used to develop a virtual sensor to infer the properties of manufactured products. The same approach is applied in Nagai and Arruda (2005) to predict the top composition of a distillation column.

It is widely reported in the literature that highly correlated variables can give numerical problems during the identification step. This is often the case for variables measured from industrial processes. This drawback can be mitigated by using projection-based techniques such as PCA and PLS, both in the linear and in the nonlinear case. The use of PCA and PLS in chemometrics applications is widely reviewed in the *IEEE Control System Magazine*, issue 5, published in 2002. A useful survey of linear and nonlinear PLS algorithms can be found in Baffi, Martin and Morris (1999).

An example of the use of PCA- and PLS-based models can be found in Flynn, Ritchie and Cregan (2005) concerning a fault detection task for a power plant. In Komulainen, Sourander and Jamsa-Jounela (2004) a review of more sophisticated techniques that can be considered as evolutions of both PCA and PLS is reported. Among the possibilities, the authors apply the dynamic PLS, which includes time-lagged values, to a fault detection task of a dearomatization process.

A comparative study of soft sensors derived using multiway PLS and an extended Kalman filter for a fed-batch fermentation process is presented in Zhang, Zouaoui and Lennox (2005). The procedure proposed allows nonlinear characteristics to be removed from the data by using suitable transformations and, hence, PLS to be adapted to a nonlinear problem.

In Liu (2005), fuzzy models are used to realize a piecewise linear time-varying model for inferring the melt index of a polyethylene process. The model is recursively updated based on PCA.

Another relevant technique proposed in the literature for variable processing is independent component analysis (ICA). It is aimed at making the variables independent, and involves higher order statistics. In Lee, Yoo and Lee (2004), ICA is used to process data relative to biological waste water treatment. An interesting comparison between ICA and PCA monitoring capabilities is reported.

In Albazzaz and Wang (2006), ICA is considered in the framework of data visualization that poses challenging problems due to the high number of variables monitored in a typical industrial plant.

Different structures can be used to model real systems. In the field of industrial applications, attention is focused on parametric structures and among these a key role is played by autoregressive models with exogenous inputs both in the linear (FIR, ARX or ARMAX) and nonlinear versions (NFIR, NARX, and NARMAX). A theoretical in-depth treatment of possible models can be found in Ljung (1999).

Regardless of the particular model structure of interest, either linear or nonlinear, a challenging task to be solved is the choice of input and output regressors, *i.e.* the choice of the model order, on the basis of measurement data and, eventually, any kind of available information. A number of contributions are available in the literature on this topic. Among them, some interesting papers will be briefly described below.

Lipschitz quotients can be used as a tool for guiding the solution of this complex task, when the assumption is made that the system nonlinearity can be represented by a smooth function in the regressors. A description of this approach can be found in He and Asada (1993) and in Nørgaard *et al.* (2000). An example of application of this method can be found in Bomberger and Seborg (1998). In the same paper, the Lipschitz quotients method is compared, but exclusively for single input–single output (SISO) systems, with a method derived from the false nearest neighbor (FNN) approach. In Nagai and Arruda (2005), the Lipschitz quotients method is used to select the lag structure of a fuzzy multiple input–single output (MISO) model of the top composition of a distillation column.

In Feil, Abonyi and Szeifert (2004), a modified version of the FNN approach, based on fuzzy clustering, is proposed to increase its efficiency. In particular, in the proposed method the model structure is estimated on the basis of the cluster covariance matrix eigenvalues. In the reference section of the paper a good selection of further work on this topic can be found. Another interesting list can be found in Lind and Ljung (2005).

A further approach that can be used for model order selection is based on input output correlation analysis (Komulainen, Sourander and Jamsa-Jounela, 2004). In Lang, Futterer and Billings (2005) a new method, derived for the correlation approach, is proposed for the identification of NARX models with input nonlinearities.

In Mendes and Billings (2001), the authors propose a method to overcome the growth in computational effort with model complexity that can compromise the search for the optimal model structure.

The ANalysis Of VAriance (ANOVA) has been proposed as a possible method for regressor selection of nonlinear models in Lind and Ljung (2005) and in Lind (2005). This approach has the valuable property of allowing the model order selection to be operated independently from the other steps required for model identification.

1.2.3 Model Identification

A good overview of nonlinear black-box structures for system identification ranging from neural networks, radial basis networks, wavelet networks, hinging hyperplanes, to fuzzy systems is reported in Sjoberg *et al.* (1995), Sjoberg, Hjalmarsson and Ljung (1994), and Juditsky *et al.* (1995). In particular, in Juditsky *et al.* (1995) the mathematical foundations of the techniques described in Sjoberg *et al.* (1995) are introduced.

The theory of identification of nonlinear systems using a polynomial model is introduced in Chen and Billings (1989), while a seminal work dealing with neural networks for identification of dynamic nonlinear systems is that of Narendra and Parthasarathy (1990). Examples of numerical applications are also given in the paper.

Beyond these classical contributions, a number of approaches are proposed in papers that mainly address the practical applications of soft sensors. The taxonomy of this huge literature represents a very hard task. Some interesting contributions are reported below, while further examples will be discussed when the literature on relevant applications is described.

A number of authors used the neuro-fuzzy approach to fuse the ease of fuzzy model interpretation with the data-driven learning capabilities of neural networks. An example can be found in Araúzo-Bravo *et al.* (2004), where soft sensors, based on neuro-fuzzy systems, named FasArt and FasBack, are used to develop dynamic adaptive models of a penicillin production process and hence to avoid human intervention.

In Luo and Shao (2006), a hybrid soft computing modeling approach in which a neuro-fuzzy system based on rough set theory and genetic algorithms is used to obtain a reduced structure neural network which estimates the freezing point of the light diesel fuel in a fluid catalytic cracking unit. The same application was considered in Yan, Shao and Wang (2004), where a different approach, based on support vector machines, is used.

Classical structures such as fuzzy c-means and fuzzy Takagi–Sugeno are used in Liu (2005) to approach the identification problem of a multivariable time-varying nonlinear system. The procedure is applied to a polyethylene process and is based on the decomposition of the nonlinear model into several subsystems.

In Rallo *et al.* (2002) a predictive fuzzy neural system and two hybrid networks, each combining a dynamic unsupervised classifier with a different kind of supervised mechanism, are applied to develop a virtual sensor to infer the properties of manufactured products.

In Nagai and Arruda (2005), a fuzzy clustering algorithm is applied to find the rule base of a soft sensor designed to infer the top composition of a distillation column.

Neural networks, namely eng-genes and multi-layer perceptons (MLPs), are designed in Li *et al.* (2005) by using genetic algorithms and are applied to the estimation of NO_x in a thermal power plant.

Identification procedures are very sensitive to the size of the data set used. Approaches tailored to alleviate negative effects of small data sets are proposed in Yan, Shao and Wang (2004) and in Feng, Chen and Tu (2005). In the last paper the problem of shortage of data is approached by using the group method of data handling to obtain the parameters of a polynomial NARMAX model. A case study referring to model radar-land-clutter reflectivity is described.

A completely different approach, based on sparse grid approximation, is proposed in Kahrs, Brendel and Marquardt (2005) to incrementally identify a nonlinear model. Interestingly, the proposed strategy allows iteration of the classical model-building steps to be reduced by a systematic strategy that combines all steps into a single algorithm.

In Park and Han (2000), a multivariate locally weighted regression (a procedure for estimating a nonlinear regression surface to data, through multivariate smoothing) for applications with high dimensionality, collinearity and nonlinearity is proposed. The approach is applied to an industrial splitter column and to a crude column.

1.2.4 Model Validation

Though performing the previous tasks could require solving also the problem of model validation, in what follows we will report on some interesting contributions to this specific topic. Generally speaking, the problem of model validation has not been solved in a definitive way. Nevertheless, in the theory of linear systems the usual approach consists of computing the autocorrelation function of the residuals and the cross-correlation functions between the residuals and the input over a set of *unseen* data (Stoica and Södersröm, 1989).

The task of nonlinear model validation is generally accomplished by extending the correlation approach to all linear and nonlinear combinations of past inputs and outputs. Classical contributions dealing with this topic are Chen, Billings and Grant (1990), Billings, Jamaluddin and Chen (1992), and Billings and Voon (1991). Two further correlation functions can be found in Mendes and Billings (2001).

Confidence levels as a tool for model validation are proposed by Papadopoulos, Edwards and Murray (2001), Dadhe and Engell (2005), and Masson *et al.* (1999). In Dadhe and Engell (2005) prediction intervals of multi-step-ahead prediction of neural networks are estimated by bootstrap methods. In Masson *et al.* (1999), K-fold cross-validation is used as a resampling technique to provide a large validation set.

1.2.5 Applications

Potential applications of soft sensors can be imagined in any industrial field. Of course some characteristics (large plants, hard-to-measure process parameters, measurement delay, working environments hostile to measuring device survival, hard control requirements, *etc.*) of the plants considered can encourage the application of this technique.

A good review of soft sensor applications in a number of different fields up to the year 2001, can be found in Dote and Ovaska (2001), while below we will mainly address the more recent applications.

Refineries are a typical application field where a wealth of contributions can be found. An early example of applications in this field is Tham, Morris and Montague (1994), where linear models are considered to solve two different modeling problems in a refinery.

A number of applications refer to the case when continuous on-line measurements are not available and models based on laboratory analysis are the only choice. In these cases, static models are mainly used (and this can affect the model quality). In Yan, Shao and Wang (2004) a model is proposed for the estimation of the freezing point of light diesel oil in distillation columns. A model for the estimation of the same quantity using a neuro-fuzzy system driven by rough set theory and genetic algorithms is proposed in Luo and Shao (2006).

An attempt to alleviate the large time intervals between lab analyses is reported in Shi *et al.* (2005) where a dynamic compensation method is proposed for the estimation of the product composition of a distillation plant.

The PLS approach is used in Park and Han (2000) to model both an industrial splitter column and an industrial crude column while in Komulainen, Sourander and Jamsa-Jounela (2004) a dynamic PLS approach is used to develop an on-line monitoring system for an industrial dearomatization process. In particular, time lags in the model are selected by using both expert plant knowledge and a classical cross-correlation analysis.

On-line fault diagnosis for a refinery fluid catalytic cracking reactor is addressed in Yang, Chen and Wang (2000), where information on the not directly accessible critical variables is extracted using a procedure based on wavelets and neural networks.

In Fortuna, Graziani and Xibilia (2005a) a dynamic nonlinear model based on a cascade of neural networks is proposed to solve the problem of the control of a debutanizer column, in the presence of large measurement delays.

Finally, a survey of typical applications, spanning from static to dynamic nonlinear neural models, that may be of interest for refineries, can be found in Fortuna, Graziani and Xibilia (2005b).

Another typical industrial application field is polymerization. A case study referring to prediction of the melt index of six low-density polyethylene (LDPE) grades produced in a tubular reactor is reported in Rallo *et al.* (2002), where a predictive fuzzy neural system and two hybrid networks, each combining a dynamic unsupervised classifier with a different kind of supervised mechanism, is used. In Xiong and Zhang (2005), dynamic neural models based on recurrent MLPs are applied to a laboratory-scale polymerization reactor. Different techniques, based on bootstrap resampling and stacked neural networks, are applied to the same lab-scale process in Zhang *et al.* (1997) and Zhang (1999).

Applications referring to fermentation processes can also be found, and a review of state estimator techniques applied to bioreactors is reported in Assis and Filho (2000). A comparative study between black-box and hybrid (gray-box) estimation methods for on-line biomass concentration estimation is reported in Janes, Legge and Budman (2002).

In Willis *et al.* (1992), an inferential controller, based on a neural network model and a classical PI controller, is used for a penicillin fermentation plant.

The pulp industry is a recurrent topic in journals dealing with industrial applications and soft sensors have been proposed in this field. In Dufour *et al.* (2005) a soft sensor to infer unmeasured variations, due to faults, in the feedstock of an industrial pulp digester are presented.

The problem of product quality in the papermaking industry, in particular the estimation of paper curl measurement, is addressed in Edwards *et al.* (1999) using neural networks. Neural models are also adopted to predict the corrosion potential in a pulp and paper mill (Bucolo *et al.* (2002)).

Emissions are of course one of the main concerns of industries, due to the restricting limits imposed by the prevailing laws. In Graziani *et al.* (2004), a static nonlinear neural model is proposed to estimate NO_x concentration in chimney emissions produced by a refinery. The proposed model overperforms an empirical model developed by the plant experts. NO_x level emissions for a power station are dealt with also in Flynn, Ritchie and Cregan (2005), where PCA and both linear and nonlinear PLS are used to synthesize a static model. The problem of NO_x emission estimation is dealt with in a novel approach, denoted eng-genes architecture, in Li *et al.* (2005).

In order to eliminate the delay time in the estimation of NO_x emissions in a power plant, an ARX soft sensor is used in Matsumura *et al.* (1998). The soft sensor is used to implement a control loop.

Another chemical matter that has become the center of attention is carbon dioxide, as its emission is recognized to be the main source of the greenhouse effect. The catalytic hydrogenation of carbon dioxide to methanol is a method for reducing its emission in the atmosphere. A hybrid structure, consisting of a mechanistic model and a neural one, to identify a methanol reactor, is proposed in Zahedi *et al.* (2005).

SO_2 and H_2S have negative effects on the phenomenon of acid rains. In Fortuna *et al.* (2003) SO_2 and H_2S in a sulfur recovery unit of a refinery are estimated using different kinds of nonlinear models.

The monitoring and control of the emission from a palm oil mill is described in Ahmad *et al.* (2004). In particular, a neural model is developed from palm oil mill data, while genetic algorithms are employed to find the optimal operating conditions to reduce the emission level.

A number of applications can be found in the literature that, even though they do not fit the classes we have introduced so far, are still of interest. In what follows, a description of these contributions is reported.

In Govindhasamy *et al.* (2005) dynamic nonlinear neural structures are proposed to model the grinding process used by a factory during the production of the aluminum disk used in disk drives. In particular, a NARX model, realized using a generalized MLP, is proposed.

In Su, Fan and Schlup (1998) the quality of epoxy/graphite fiber composites, measured using the degree of cure (DOC), are estimated using a recurrent neural network. This network is the core of a soft sensor that manipulates the outputs of a dual heat-flux sensor to estimate the DOC value.

Lin *et al.* (2005) describe the steps to be performed in a typical soft sensor design and apply the proposed techniques to the real-time estimation of the behavior of a cement kiln.

Ruan *et al.*'s (2003) contribution is the description of somewhat embryonic results, further improved in Roverso and Ruan (2004). Both these papers deal with the possibility of using a number of artificial intelligence techniques, including noise analysis, neural networks, fuzzy theory, and wavelets, to estimate feedwater flow in a nuclear plant. In particular, classical measuring techniques, *i.e.* Venturi flowmeters, are integrated by using the proposed techniques to realize a novel soft sensor.

Food factories are another interesting field for the application of soft sensors. In Wang, Chessari and Karpiel (2001) the problem of product quality control in a food cooking extruder is addressed. An on-line dynamic model between the influential variables and the product quality attributes is proposed to design a feedback control.

In Janes, Yang and Hacker (2005), the authors focus their attention on the problem of pork farm odor modeling that can be of great interest to mitigate negative effects for the residents of rural areas. In particular, multiple-components multiple-factor neural models are compared to linear models and the superiority of the neural approach is shown.

In Englund and Verikas (2005), the problem of outlier detection in an offset lithographic printing process is addressed, and a structure containing both a fuzzy system for data filtering and artificial neural networks is proposed.

2

Virtual Instruments and Soft Sensors

2.1 Virtual Instruments

In this chapter, the concepts of virtual instruments (VIs) and of soft sensors will be introduced in some detail. In particular, we will commence in this section with an introduction to VIs, to focus in the next one on soft sensors, that are the main topic of the book.

VIs can be considered as a wider class than soft sensors. In fact, soft sensors focus on the process of estimation of any system variable or product quality by using mathematical models, substituting some physical sensors and using data acquired from some other available ones.

For their part, VIs are based on software that performs any of the typical actions involved in a measurement and/or control problem, by exploiting available instrumentation, computers and software. This action can either involve, or not, modeling capabilities typical of soft sensors.

VIs are the result of the rapid diffusion that has taken place in the last 20 years of low-cost Windows-based personal computers, Macs, and workstations in any engineering application field, along with performing software (Foster ,1998).

They represent an alternative paradigm to traditional instruments and allow us both to customize the measuring facility capabilities to user application and to take control of the way measurement results are used and presented.

The flexibility of VIs is obtained by software that is used to transform a collection of facilities into the customized instrumentation suitable for the measurement task of interest. Also, the use of software allows the adaptation, at different times, of the available resources to new measuring problems in such a way as to adapt the measuring system to new scenarios, and hence to better exploit available resources.

Three main functional blocks can be recognized in any instrument, and all of them have been affected by changes introduced by the introduction of VIs (Combs, 1999):

- measurement;
- computation;
- user interface.

As regards measurement, this is the very first action that a measuring system performs on the observed variable, in order to extract some kind of meaningful signal. Signals extracted from real plants are analog in nature, while the majority of modern instrumentation is digital. An analog to digital (A/D) converter is, therefore, required at this stage along with conditioning circuitry, to adapt real-world signals to the A/D input span.

At this stage, VIs are used mainly to the ease the setting and control of measuring systems, especially in routine measurement surveys that require a large number of actions, and are usefully automatized using adequate software. This is the typical application when a VI is designed, and used, to drive a stand-alone system.

Data acquired from the measurement hardware do not necessarily correspond to the searched information. It is generally required either to filter data in some way, to combine data acquired from the same device in different times or different points, or to combine data acquired from different measuring facilities. At this level, computation capability plays a key role and it is possible either to have instruments with local computation capabilities or to send raw data to some kind of intelligent system that performs the required data manipulation. Of course, these two solutions impose different constraints on the designer and have conflicting performances. The first one is generally more expensive, because of the need to have a number of local intelligent systems, while the other will need a faster communication system to handle the large quantity of raw data that needs to be transferred.

The computation level is of course one of the most characterizing aspects of VIs. In fact, by using adequate software it is possible to use general measuring devices, *e.g.* acquisition cards or modular instrumentation, to acquire data and combine them in a virtually infinite number of ways. Moreover, the same devices can be re-used in different applications simply by changing the algorithms used.

Though modern electronic instrumentation is totally different from older analog measuring systems, producers try to maintain the traditional look of the instruments. This is because users experience great difficulty when the objects they are used to change their appearance. As an example, a modern digital oscilloscope has an internal structure that is totally different from traditional analog ones. Nevertheless, both of them look very similar and present a number of input channels, some knobs and buttons to control the instrument operations, and a screen. This allows the user to change from older systems to newer ones with a minimum of difficulty: only if he is interested, will he search at some convenient time for new functionalities.

This general rule has been maintained in the case of VIs. The measuring system might be realized by using traditional stand-alone instrumentation, with its own user interface, or by using modular instrumentation; in that case the user interface is almost totally missing, and so the VI designer will produce, on a host PC monitor, a user interface that mimics the traditional instrumentation front panel and that will allow interaction with the measuring system. In this way, the user will find digital knobs or buttons to control the measurement system and will observe the information in which he is interested on some kind of graph, indicator, or in the case of surveillance systems, for example, a simulated LED will be turned on by the software together with some kind of acoustic signaling.

An example of a front panel of a VI is shown in Figure 2.1.

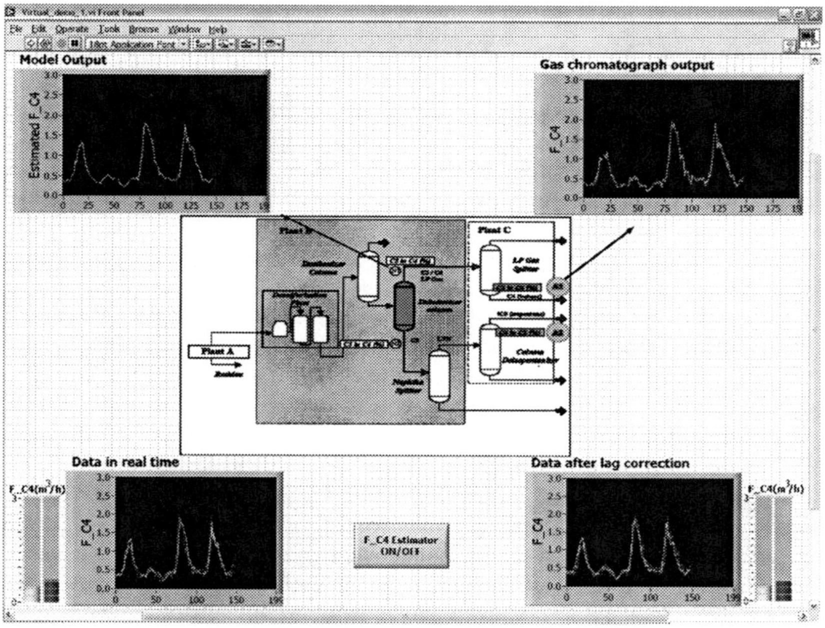

Figure 2.1. An example of a VI front panel

Figure 2.1 shows the front panel of an instrument designed for the estimation of the products of a debutanizer column that will be described in greater detail elsewhere in the book. In this case, the core of the VI was a soft sensor based on a cascade of neural network dynamic models whose objective was the prediction of products concentration without the large delay introduced by traditional measuring systems. In particular, in Figure 2.1 the presence of a number of time plots, a button, and some indicators can be recognized. The button was introduced to allow the user to turn the VI on and off, while both the time graphs and the indicators show the instrument outputs.

The availability of low-cost computers with programming capabilities has produced a far-reaching and rapid evolution also in the possible measuring hardware configurations. In fact, on the one hand, modern instrumentation is configured as a digital system whose core is a microprocessor that controls all actions required to perform measurements, while on the other one, it is more and more common that the instruments communication capabilities allow for the realization of distributed measuring networks where a number of devices cooperate and exchange relevant information, either using a shared standard or custom communication protocol.

The simplest and most widely used class of measuring instruments are stand-alone devices. In this case, a single box contains all the resources required to perform measurements and is equipped with a front panel that is used both to set up the instrumentation and to display measurement results. VIs in this case are used to realize virtual front panels on a PC that mimics the hardware front panel. These instruments can operate by themselves and eventually, if they have digital communication capabilities, can be used in a multiple instrument system. For example, this is the typical configuration used in IEEE 488.2 compliant measurement systems.

With the evolution of computers, modular measuring instrumentation became available. In this case, the instrument front panel is totally missing and the only available option to use modular systems is to insert them into a frame and to program them by software. VIs are widely used in this context to program and use measuring stations.

Generally, modular measuring systems are realized by using some standard, such as VME, VXI or PXI.

They greatly outperform traditional stand-alone systems especially when the required system throughput is high, but they are quite expensive. Also, since they are mainly software defined, the available resources can be reconfigured virtually an infinite number of times and this is a valuable possibility both for R&D purposes, when the measured variables change frequently, and for maintenance and control applications, when measuring systems are often updated.

The latest evolution of modern electronic instrumentation is based on networked devices. In this case, each instrument acts as a computer, capable of being connected on some LAN and of sharing a common communications line or wireless link within a small geographic area, or even by using the Internet. These last scenarios are typical of monitoring and control in industrial applications due to the distributed nature of processes involved and of air quality monitoring in large urban and industrial areas, where a number of measuring stations are installed at adequately chosen points to obtain information about air quality.

Of course, also in the latter case, VIs can be very useful especially if software tools with networking programming capabilities are used, which can greatly simplify both the troubleshooting and running of the measuring system.

As mentioned before, the success of Vis is mainly due to the possibility of reconfiguring them to perform custom functions. In this sense, they look quite different from traditional instrumentation, designed for one specific measurement task.

Apart from the initial system design, which for VIs must start with the identification of some minimum hardware resources, the design of a VI is a matter of software programming. The importance of having access to powerful programming languages is therefore fundamental. It is by using programming languages that general measuring tools are customized by the user to the intended application. In this sense, the VI designer has much more freedom than the traditional instrumentation designers, even if such flexibility can pose serious problems of re-using available software, unless close attention is paid to realizing modular software in which previously designed software procedures can be included for new applications.

There are two categories of development environments for VIs available for the designer:

- textual languages;
- graphical languages.

In the case of textual languages, traditional programming languages, such as BASIC or C/C++, are used to realize a VI. Generally, programming in this case involves the use of functions specifically developed by the vendor for the hardware in question, and that govern the specific instrument functionalities and I/O functions for communication purposes.

Graphical programming allows the design of VIs using functional blocks (icons) that perform specific tasks. These blocks perform desired actions, from simple to very complex ones. Information is transferred from one icon to another by suitably wiring them. Also, programming languages come with a number of libraries (*e.g.* to perform signal spectrum analysis or digital signal processing) that greatly improve the language potentialities.

To better explain how graphical programming languages work, we will refer to the widely used LabVIEW™ by National Instruments, which can be considered a standard *de facto* for VI design, though competing languages have been proposed by other companies and still exist on the market.

A VI consists of a front panel that mimics the user interface and a block diagram that lies at the back of the front panel and is used to graphically define the VI functionalities. Moreover, during the design phase of a VI, other available VIs can be used as subroutines to perform simpler tasks in a hierarchical way.

The front panel of a VI, presented on the computer monitor, represents the user interface and intentionally looks very like a traditional instrument panel. This is the part of the VI that allows the user to act on the instrument, and eventually on the connected measuring hardware, by setting measurement parameters and loading data files. To obtain this capability, the designer can use a number of knobs, buttons, switches and so forth. Also, the front panel is used to show the user the results of VI operations (including measurement and computation results).

A number of tools are available to show VI outputs in the best way. The designer can, in fact, use time plots, XY graphs, numeric indicators, to give just a few examples, or even digital LEDs in those cases when an immediate alarm is better suited to the application.

As an example, in Figure 2.2 a very simple front panel showing a time graph, a numerical indicator, and a LED is reported.

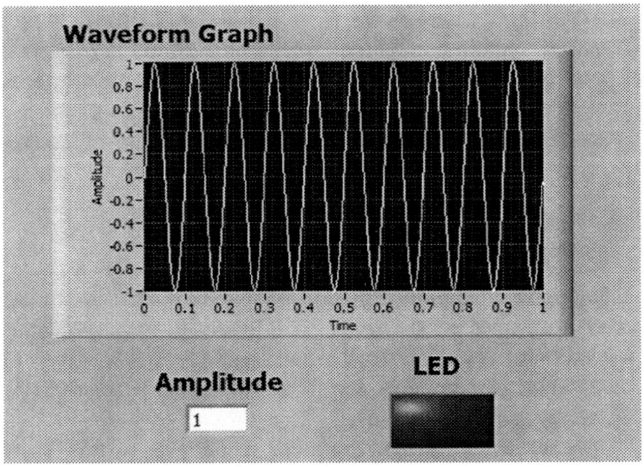

Figure 2.2. Example of a typical VI front panel

The block diagram is the core of a VI: the designer assembles here a number of available functional blocks, in the form of icons that carry out specific actions on input data, and produce corresponding output results. These blocks can either be part of the graphical programming language or have been realized by the user during the development of previous VIs. The function of each block can be simple, such as adding two input variables and giving the resulting sum, or very complex, such as performing sophisticated statistical analysis of input vectors.

Data are passed from one block to the next one using software wires. In the same way, elements in the front panel have icons in the block diagram that can be wired to functional blocks in the block diagram to allow commands from the user to be passed to the graphical software and final results to be shown on the front panel.

Figure 2.3 shows an example of a VI block diagram. The VI accepts numerical inputs from the VI front panel and returns their sum on a front panel indicator.

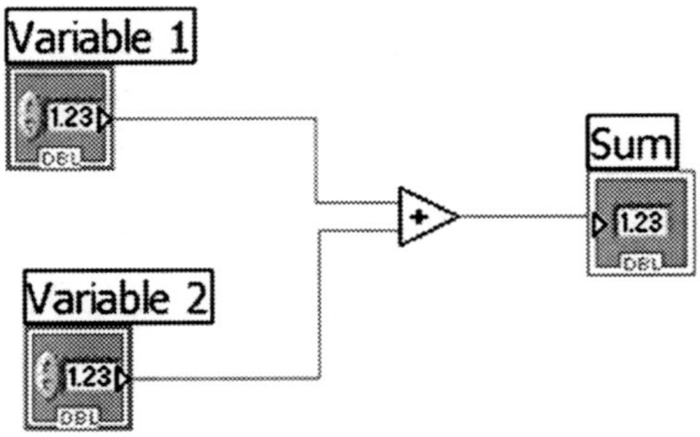

Figure 2.3. Examples of a typical VI block panel

Finally, Figure 2.4 shows a less didactic example of a VI block diagram. The reported example is part of the block diagram of the VI whose front panel is given in Figure 2.1, which was developed for the realization of a soft sensor for a debutanizer column.

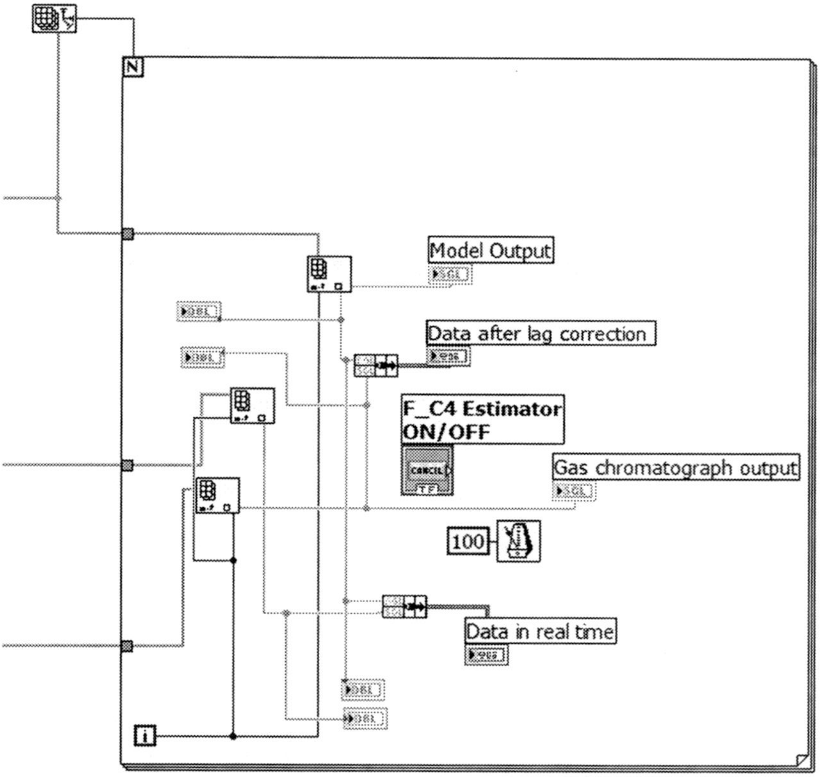

Figure 2.4. Part of a block diagram of the VI reported in Figure 2.1, for estimating debutanizer products

It is worth noticing that in the block diagram powerful tools for instrument driving, supplied by vendors, networking facilities, and data storage are available, all of which make VIs very flexible devices.

Some functional blocks can take inputs from an A/D converter and/or hardware measuring device and elaborate them on the basis of a user designed code. In other words, a block in the block diagram can be a soft sensor. Also, the soft sensor can either be designed using the graphical programming language that implements the VI or can be software, developed using a textual programming language, depending on the most suitable approach.

As a final remark on the difference between textual and graphical programming languages for VI design, it should be noted that, though graphical languages are more suitable for their eye appeal in presenting measurement results to the end user, they generally are very resource demanding and especially in real-time applications can introduce an unacceptable delay. In contrast, textual based VIs are much more conservative as regards computing resources and can be very efficient tools for real-time control applications. As usual, the final choice between the two alternatives will depend on the designer, who will need to take into account velocity constraints imposed by the application.

Of course, hybrid solutions where complex data elaboration is performed using textual programming languages, and where the user interface is realized using a graphical language, can be developed.

2.2 Applications of Soft Sensors

There are a number of reasons why soft sensors can be profitably used in industrial applications; currently they are becoming routine tools with the trend moving from open-loop information tools for the operator towards sensors in closed-loop inferential and/or adaptive control schemes.

Moreover, the wide availability of on-line analyzers and digital systems that are used both for monitoring and control give designers and operators the tools required for the design and implementation of soft sensors with a minimum, or even null, increase in the initial costs. In what follows, a description of typical soft sensors applications is given.

2.2.1 Back-up of Measuring Devices

A huge number of measuring devices connected to realize distributed monitoring networks, are used from industrial plants for monitoring and control pourposes. They routinely acquire a very large quantity of data. In fact, monitoring the state of a plant, even at a fixed time instant, could require sampling hundreds or even thousands of different variables.

Such measuring devices, and the corresponding data transmission systems, are required to face very harsh working environments. It is not surprising that working conditions impose both the use of very robust measuring hardware and periodic maintenance procedures. Notwithstanding such precautions, faults in measuring devices occur. Faults can come either in the form of abrupt changes in the working mode of measuring devices or in the form of slow changes of metrological characteristics. The latter can be even more dangerous than the former, because it is more difficult to detect and can hence cause malfunctioning of control systems.

Irrespective of whether a maintenance intervention be programmed or accidental, the measuring hardware needs to be turned off and suitably substituted. The back-up of measuring instrumentation is a typical application of soft sensors: an inferential model is in this case specifically designed to momentarily substitute unavailable measuring equipment and to avoid degradation of plant performance and rises in cost.

This particular scenario imposes restrictions on the soft sensor designer's possible choices. In particular, care must be taken in those cases when the variable inferred by the soft sensor is the output of a dynamic system. In fact, two possible choices are common:

- to restrict the model structure to moving average (MA) or nonlinear moving average (NMA) models that do not require past samples of the output variable. This corresponds to limiting the class of possible models to the class of finite impulse response (FIR) structure (including their nonlinear generalization), eventually decreasing the model performance with respect to more general model structures;
- to use autoregressive models as infinite-step-ahead predictors. In this case, the model has among its inputs past samples of its own estimations with corresponding feedback of model errors. Such structures are generally more efficient than the corresponding MA or NMA structures in the very first predicting steps but, generally, their performance quickly degrades due to error propagation. This is very true when the envisaged maintenance interval is very large compared to the system dynamics, so that a large number of successive samples are required to be estimated.

2.2.2 Reducing the Measuring Hardware Requirements

Using a software tool instead of a measuring hardware device represents, of course, a source of possible budget saving. Experts can therefore be encouraged to design inferential models that are intended to definitively substitute hardware devices, which become available for further reallocation.

Also in this case, a NMA model should be preferred to autoregressive structures. In any case much care should be taken to critically analyze model performance, due to the lack of any redundancy, and periodic model validation should be performed by temporarily inserting measuring devices and eventually proceeding to soft sensor retuning.

The problem of periodic soft sensors validation and eventual retuning is actually a common issue for any application of soft sensors. The need for such retuning can be due to a change in a new process working point (not considered during the soft sensor design phase), which can be detected by critically analyzing system inputs. This analysis should be constantly performed by checking for violation of suitable thresholds imposed for each input variable.

Soft sensor retuning is also needed when a change in system parameters occurs, in the case of slowly time-varying systems (*e.g.* due to seasonal variations).

The application scenario considered in this subsection is particularly sensitive to these problems. In fact, in the other applications described in this section, measuring hardware is always available (at least after a finite maintenance period) and this allows the required soft sensor validation operations to be accomplished. When the designer intends to eliminate measuring hardware, the availability of measuring facilities for sensor validation must be suitably planned, and extra hardware must be used temporarily.

2.2.3 Real-time Estimation for Monitoring and Control

The real-time estimation of system variables obtained using soft sensors, as opposed to its delayed measurement by means of hardware measuring devices, represents the most valuable feature of soft sensors; this is due both to the possibility to design a very efficient soft sensor and to the importance of the corresponding benefits in terms of process performance.

Any measuring instrument requires a finite time to perform the actions needed to give the final variable measurement. Though such a time can be very small in a number of applications, in some cases it can be significant. To give an example, this is the case with some gas chromatographs that require measuring times of the order of minutes or even greater. Moreover, due to the high cost of some measuring devices used in industrial applications, variables can sometimes be inferred on the basis of data acquired using measuring hardware that can be located on different processes, with a corresponding further delay (see, for example, the application reported in Chapter 8). Should this time be comparable with system dynamics, the measuring time can be a significant source of delay.

In the case of measuring instrumentation used for monitoring purposes, this corresponds to a delay in the time in which data are presented to the operator, with no relevant consequences, unless this information is important for safety issues.

When information about a variable value is needed in a closed-loop control scheme the effects of delay can decrease system performance to the point that the measuring hardware is not suitable for the control application.

In this class of applications the variable measurement is always available, albeit with a relevant delay. This allows the use of Auto-Regressive with eXogenous inputs (ARX) or Nonlinear ARX (NARX) model structures, which perform finite (and small) step-ahead prediction of the variable.

The real-time estimation obtained by the soft sensor can be used by the controller, while the corresponding delayed measurements allow the soft sensor performance to be improved, by avoiding the error propagation effect mentioned in the previous subsection.

2.2.4 Sensor Validation, Fault Detection and Diagnosis

An industrial control system can be seen as a hierarchy of at least three levels: the first level is the control level, which implements the actual control loop by means of feedback and feedforward controllers, state observers, parameter estimators, and so on. Above the control level, the supervision level accomplishes the task of continuously monitoring the operational life of the process, making the process operation almost independent from the presence of human operators. The highest level is dedicated to management, coordination and optimization activities, which provide the control system with high-level directives in order to maximize the performance of the system with respect to certain criteria.

Fault detection and diagnosis are part of the supervision functions accomplished by modern industrial control systems. In the past, the supervision function was essentially limited to checking important variables, and the consequent raising of alarms if some safety thresholds were trespassed. This was actually an early stage of fault detection. On the other hand, at present, fault detection and diagnosis is performed by means of advanced techniques of mathematical modeling, signal processing, identification methods, computational intelligence, approximate reasoning, and many others. The main goals of modern fault detection and diagnosis systems are to:

- perform early detection of faults in the various components of the system, possibly providing as much information as possible about the fault which has occurred (or is occurring), like size, time, location, evaluation of its effects;
- provide a decision support system for scheduled, preventive, or predictive maintenance and repair;
- provide a basis for the development of fault-tolerant systems.

Fault detection and diagnosis strategies always exploit some form of redundancy. This is the capability of having two or more ways to determine some characteristic properties (variables, parameters, symptoms) of the process, in order to exploit more information sources for an effective detection and diagnosis action. The main idea underlying all fault detection strategies is to compare information collected from the system to be monitored with the corresponding information from a redundant source. A fault is generally detected if the system and the redundant source provide two different sets of information. There can be three main kinds of redundancy: physical redundancy, which consists of physically replicating the component to be monitored; analytical redundancy, in which the redundant source is a mathematical model of the component; knowledge redundancy, in which the redundant source consists of heuristic information about the process. When dealing with industrial applications, an effective fault detection and diagnosis algorithm must usually exploit a combination of redundancy sources, rather than a single one.

Sensor validation is a particular kind of fault detection, in which the system to be monitored is a sensor (or a set of sensors). At a basic level, the aim of sensor validation is to provide the users of a measurement system (that can be human operators, measurement databases, other processes, control systems, *etc.*) with an evaluation about the reliability of the measurement performed. At a higher level, a sensor validation system may also provide an estimate of the measurement in the case in which the actual sensor is out of order. In this framework, soft sensors are a valuable tool to perform sensor validation. Their usefulness is twofold. First, they can be exploited as a source of analytical redundancy. They can in fact be paralleled with actual sensors, and faults can be detected by comparison between the outputs of actual and soft sensors. Second, they can be exploited to provide an estimate of the sensor output in the case of sensor fault. Therefore, they can be used as a back-up device once a fault has been detected.

2.2.5 What-if Analysis

The design process of control systems requires the process behavior to be described via adequate theoretical/data-driven models that might be able to predict the system output corresponding to suitable input trends, for a given time span.

A model is used in this case to perform simulation of the system dynamics corresponding to input trends that are of interest, with the aim of obtaining both a deeper understanding of system behavior and/or designing suitable control policies. This particular use of process models to perform simulation is called what-if analysis.

Though first principle models could be a better choice due to their capability of describing the phenomena ruling the process, the difficulty of obtaining accurate enough models in a reasonable time can lead experts to adopt data-driven inferential models.

In the case of what-if analysis, inputs are therefore synthetic quantities, *i.e.* they are designed in order to analyze system reactions on a time span that makes sense, in accordance with system dynamics.

In this case, NARX models can be a suitable choice, due to the finite time span used in simulations. In fact, in this way, model error effects propagate only for a small number of iterations that must, however, be carefully fixed by the designer. It is also worth noting that, in the case of what-if analysis, input variables are noise-free, thus improving simulation performances.

On the other hand, much attention must be addressed to a careful choice of input trends. Much more than in the cases described in previous subsections, data used during soft sensor design must represent the whole system dynamics.

Also, the usual model validation should be followed by a further test phase in which canonical signals are used to force the real plant, and recorded plant reactions are compared to model simulations. A case study describing the design of a soft sensor to perform the what-if analysis of a real process will be reported in Chapter 8.

3

Soft Sensor Design

3.1 Introduction

This chapter gives a brief description of the methodologies used in this book for soft sensor design. It is intended to help the reader in understanding the approach used in the following chapters and not to give an exhaustive treatment of theoretical topics relevant to soft sensors: readers interested in a deeper description of theoretical aspect can refer to the cited bibliography.

The chapter is organized following the typical steps that a soft sensor designer is faced with. As reported in previous chapters, soft sensors are mathematical models that allow us to infer relevant variables on the basis of their dependence on a set of influential variables. In line with the topic of the book only data-driven soft sensor design techniques will be considered in this chapter.

The methodologies described will be reconsidered in the following chapters using a number of suitable case studies. All the applications considered were developed using data taken from plant databases of real industrial applications, with only the preliminary manipulation of data scaling when required for reasons of confidentiality.

3.2 The Identification Procedure

The soft sensor design based on data-driven approaches follows the block scheme reported in Figure 3.1. A number of constraints, when using this scheme, depends on the objective for which the soft sensor is required. As an example, a soft sensor designed for measuring hardware back-up cannot use past measured samples of the inferred plant variable. This consideration will impose contrains in the block "Model structrure and regressor selection". As a second example, if the soft sensor will be designed to reduce the effect of measurement delays in a closed loop control scheme different constraints should be considered for the same block. In fact, past samples of the inferred variables could be available, suggesting for using them in the model. At the same time, high model prediction capabilities are mandatory.

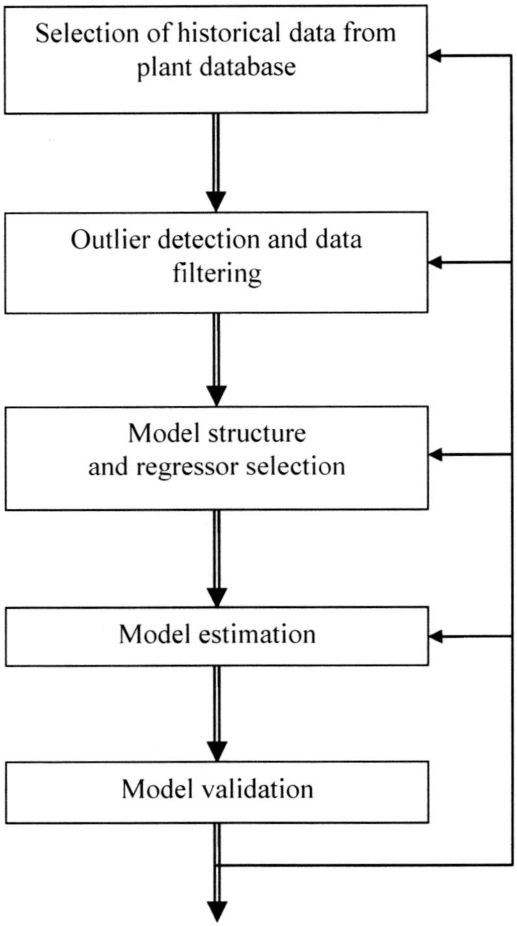

Figure 3.1. Block scheme of the identification procedure of a soft sensor

As regards the first block reported in Figure 3.1, a preliminary remark is needed. Generally, the first phase of any identification procedure should be the experiment design, with a careful choice of input signals used to force the process (Ljung, 1999). Here this aspect is not considered because the input signals are necessarily taken from the historical system database. In fact, due to questions of economy and/or safety, industries can seldom (and sometimes simply cannot) perform measurement surveys.

This poses a number of challenging problems for the designer, such as: missing data, collinearity, noise, poor representativeness of system dynamics (an industrial system spends most of its time in steady state conditions and little information on system dynamics can be extracted from data), *etc.*. A partial solution to these problems is the careful investigation of very lengthy records (even of several years) in order to find relevant data trends.

In this phase, the importance of interviews with plant experts and/or operators cannot be stressed enough. In fact, they can give insight into relevant variables, system order, delays, sampling time, operating range, nonlinearity, and so forth. Without any expert help or physical insight, a soft sensor design can become an unaffordable task and data can be only partially exploited.

Moreover, data collinearity and the presence of outliers need to be addressed by applying adequate techniques, as will be shown in the following chapters of the book.

Model structure is a set of candidate models among which the model should be searched for. The model structure selection step is strongly influenced by the purpose of the soft sensor design for a number of reasons. If a rough model is required or the process works close to a steady state condition, a linear model can be the most straightforward choice, due to the greater simplicity of the design phase. A linear model can also be the correct choice when the soft sensor is to be used to apply a classical control strategy. In all other cases a nonlinear model can be the best choice to model industrial systems, which are very often nonlinear.

Other considerations about the dependence of the model structure on the intended application have already been reported in Chapter 2.

Regressor selection is closely connected with the problem of model structure selection. This aspect has been widely addressed in the literature in the case of linear models. In this chapter, a number of methods that can be useful also for the case of nonlinear models will be briefly described.

The same consideration holds true for model identification, consisting in determining a set of parameters which will identify a particular model in the selected class of candidates, on the basis of available data and suitable criteria. In fact, approaches such as least mean square (LMS) based methodologies are widely used for linear systems.

Though a corresponding well established set of theoretical results is not available for nonlinear systems, methodologies like neural networks and neuro-fuzzy systems are becoming standard tools, due to the good performance obtained for a large number of real-world applications and the availability of software tools that can help the designer.

In the applications described in this book we mainly use multi-layer perceptron (MLP) neural networks. The topic of neural network design and learning is beyond the scope of this book. Interested readers can refer to Haykin (1999).

The last step reported in Figure 3.1 is model validation. This is a fundamental phase for data-driven models: a model that fits the data used for model identification very well could give very poor results in simulations performed using new sets of data. Moreover, models that look similar according to the set of available data can behave very differently when new data are processed, *i.e.* during a lengthy on-line validation phase.

Criteria used for model validation generally depend on some kind of analysis performed on model residuals and are different for linear and nonlinear models. A number of validation criteria will be described later in this chapter and will be applied to case studies in the following chapters.

Finally, it should be borne in mind that the procedure shown in Figure 3.1 is a trial and error one, so that if a model fails the validation phase, the designer should critically reconsider all aspects of the adopted design strategy and restart the procedure trying different choices. This can require the designer going back to any of the steps illustrated in Figure 3.1, and using all available insight until the success of the validation phase indicates that the procedure can stop.

3.3 Data Selection and Filtering

The very first step in any model identification is the critical analysis of available data from the plant database in order to select both candidate influential variables and events carrying information about system dynamics, relevant to the intended soft sensor objective. This task requires, of course, the cooperation of soft sensor designer and plant experts, in the form of meetings and interviews. In any case, a rule of thumb is that a candidate variable and/or data record can be eliminated during the design process, so that it is better to be conservative during the initial phase. In fact, if a variable carrying useful information is eliminated during this preliminary phase, unsuccessful iteration of the design procedure in Figure 3.1 will occur with a consequent waste of time and resources.

Data collection is a fundamental issue and the model designer might select data that represent the whole system dynamic, when this is possible by running suitable experiments on the plant. High-frequency disturbances should also be removed.

Moreover, careful investigation of the available data is required in order to detect either missing data or outliers, due to faults in measuring or transmission devices or to unusual disturbances. In particular, as in any data-driven procedure, outliers can have an unwanted effect on model quality. Some of these aspects will now be described in greater detail.

Data recorded in plant databases come from a sampling process of analog signals, and plant technologists generally use conservative criteria in fixing the sampling process characteristics. The availability of large memory resources leads them to use a sampling time that is much shorter than that required to respect the Shannon sampling theorem. In such cases, data resampling can be useful both to avoid managing huge data sets and, even more important, to reduce data collinearity.

A case when this condition can fail is when slow measuring devices are used to measure a system variable, such as in the case of gas chromatographs or off-line laboratory analysis. In such cases, static models are generally used. Nevertheless, a dynamic MA or NMA model can be attempted, if input variables are sampled correctly, by using the sparse available data over a large time span. Anyway, care must be taken in the evaluation of model performance.

Digital data filtering is needed to remove high-frequency noise, offsets, and seasonal effects.

Data in plant databases have different magnitudes, depending on the units adopted and on the nature of the process. This can cause larger magnitude variables to be dominant over smaller ones during the identification process. Data scaling is therefore needed. Two common scaling methods are min–max normalization and z-score normalization. Min–max normalization is given by:

$$x' = \frac{x - min_x}{max_x - min_x}\left(max_{x'} - min_{x'}\right) + min_{x'} \tag{3.1}$$

where:

x is the unscaled variable;
x' is the scaled variable;
min_x is the minimum value of the unscaled variable;
max_x is the maximum value of the unscaled variable;
$min_{x'}$ is the minimum value of the scaled variable;
$max_{x'}$ is the maximum value of the scaled variable.

The z-score normalization is given by:

$$x' = \frac{x - mean_x}{\sigma_x} \tag{3.2}$$

where:

$mean_x$ is the estimation of the mean value of the unscaled variable;
σ_x is the estimated standard deviation of the unscaled variable.

The z-score normalization is preferred when large outliers are suspected because it is less sensitive to their presence.

Data collected in plant database are generally corrupted by the presence of outliers, *i.e.* data inconsistent with the majority of recorded data, that can greatly affect the performance of data-driven soft sensor design. Care should be taken when applying the definition given above: unusual data can represent infrequent yet important dynamics. So, after any automatic procedure has suggested a list of outliers, careful screening of candidate outliers should be performed with the help of a plant expert to avoid removing precious information. Data screening reduces the risk of outlier masking, *i.e.* the case when an outlier is classified as a normal sample, and of outlier swamping, *i.e.* the case when a valid sample is classified as an outlier.

Outliers can either be isolated or appear in groups, even with regular timing. Isolated outliers are generally interpolated, but interpolation is meaningless when groups of consecutive outliers are detected. In such a case, they need to be removed and the original data set should be divided into blocks to maintain the correct time sequence among data, which is needed to correctly identify dynamic models. Of course, this is not the case with static models, which require only the corresponding samples for the remaining variables to be removed.

The first step towards outlier filtering consists in identification of data automatically labeled with some kind of invalidation tag (*e.g.* NaN, Data_not_Valid, and Out_of_Range). After this procedure has been performed, some kind of detection procedure can be applied. Though a generally accepted criterion does not exist, a number of commonly used strategies will be described. In particular, the following detection criteria will be addressed:

- 3σ edit rule;
- Jolliffe parameters;
- residual analysis of linear regression.

In the 3σ edit rule, the normalized distance d_i of each sample from the estimated mean is computed:

$$d_i = \frac{x_i - mean_x}{\sigma_x} \tag{3.3}$$

and data are assumed to follow a normal distribution, so that the probability that the absolute value of d_i is greater than 3 is about 0.27% and an observation x_i is considered an outlier when $|d_i|$ is grater than this threshold.

To reduce the influence of multiple outliers in estimating the mean and standard deviation of the variable, the mean can be replaced with the median and the standard deviation with the median absolute deviation from the median (MAD). The 3σ edit rule with such a robust scaling is commonly referred to as the Hampel identifier. Other robust approaches for outlier detection are reviewed in Chiang, Perl and Seasholtz (2003).

The Jolliffe method, reviewed in Warne *et al.* (2004), is based on the use of the following three parameters, named d_{1i}^2, d_{2i}^2, d_{3i}^2, computed on the variables z, obtained by applying either the principal component analysis (PCA) or projection to latent structures (PLS) to the model variables. The parameters are computed as follows:

$$d_{1i}^2 = \sum_{k=p-q+1}^{p} z_{ik}^2 \tag{3.4}$$

$$d_{2i}^2 = \sum_{k=p-q+1}^{p} \frac{z_{ik}^2}{l_k} \tag{3.5}$$

$$d_{3i}^2 = \sum_{k=p-q+1}^{p} z_{ik}^2 \, l_k \tag{3.6}$$

where:

index i refers to the ith sample of the considered projected variable;

p is the number of inputs;
q is the number of principal components (or latent variables) whose variance is less than one;
z_{ik} is the ith sample of the kth principal component (or latent variable);
l_k is the variance of the kth component.

Statistics in Equations 3.4 and 3.5 have been introduced to detect observations that do not conform with the correlation structure of the data. Statistic 3.6 was introduced to detect observations that inflate the variance of the data set (Warne *et al.*, 2004).

Suitable limits to any of the three statistics introduced above can be used as a criterion to detect outliers. PCA and PLS can also be used directly to detect outliers by plotting the first component *vs.* the second one and searching for data that lie outside a specified region of the plot (Chiang, Perl and Seasholtz, 2003).

A final technique considered here is the residual analysis of linear correlation. This is based on the use of a multiple linear regression between dependent and independent variables in the form:

$$y = X\,\beta + \varepsilon \tag{3.7}$$

where:

y is the vector of the system output data;
X is a matrix collecting input variable data;
β is a vector of parameters;
ε is a vector of residuals.

The procedure requires the least square method to be applied to obtain an estimation of β:

$$\hat{\beta} = (X^T\,X)^{-1}X^T y \tag{3.8}$$

so that the estimated output is

$$\hat{y} = X\,\hat{\beta} \tag{3.9}$$

and the model residual can be computed as

$$r = y - \hat{y} \tag{3.10}$$

The residuals are plotted together with the corresponding 95% confidence interval (or any other suitable interval). Data whose confidence interval does not cross the zero axis are considered outliers. As an example, in Figure 3.2 the results of a case study described in Chapter 4 (Figure 4.21) are reported.

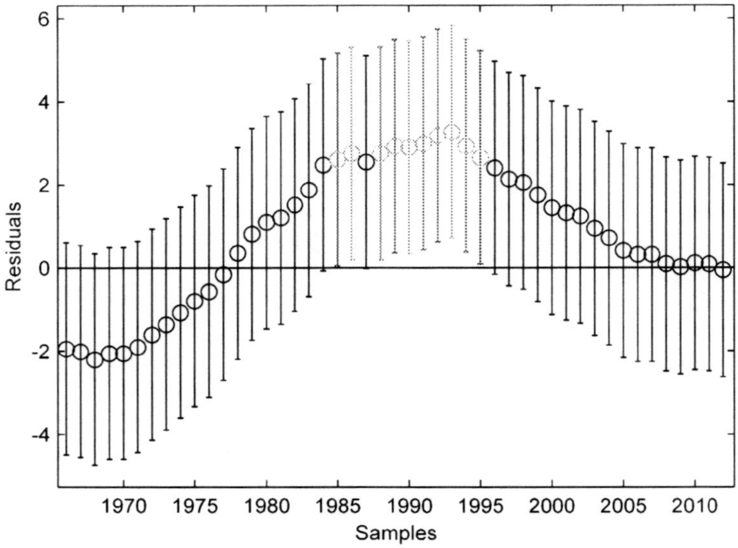

Figure 3.2. An example of outliers detected using the linear regression technique: outliers correspond to segments that do not cross the zero line and are reported in gray

Nonlinear extensions of techniques for outlier detection introduced so far can be used. Examples are PLS, which can be replaced with nonlinear PLS (NPLS), and linear regression, which can be substituted with any kind of nonlinear regression.

As a final remark, it should be noted that outlier search methods use very simple models (*e.g.* only static models are considered for the case of linear regression) between inputs and outputs, and suggest as outliers all data that do not fit the model used with a suitable criterion. This implies that the information obtained needs to be considered very carefully. In fact, automatic search algorithms tend to label as outliers everything that does not fit the rough model used. This can lead to the elimination of data that carry very important information about system dynamics and can significantly affect the results of the procedure used for soft sensor design.

The final choice about data to be considered as outliers should be performed by a human operator, with the help of plant experts.

3.4 Model Structures and Regressor Selection

Here some general model structures to be used for data-driven models will be introduced. In particular, we will start with linear models and then generalize about the corresponding nonlinear models.

Whatever the model structure, the very first representation of a system is that of an oriented structure, where a set of dependent variables, *i.e.* the system outputs, are the consequence of a set of independent variables, *i.e.* the system inputs. This schematization of a system model is reported in Figure 3.3.

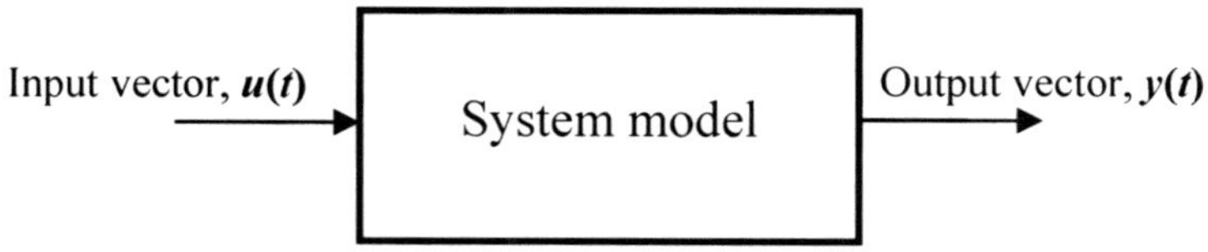

Figure 3.3. Scheme of an oriented system

The general model of a linear system is

$$y(t) = G(z^{-1})u(t) + H(z^{-1})e(t) \tag{3.11}$$

where, for Single Input–Single Output (SISO) systems, $G(\cdot)$ and $H(\cdot)$ are transfer functions, z^{-1} is the time delay operator and $e(t)$ is a white noise signal, with a corresponding probability density function; and Equation 3.11 can then be rewritten as

$$A(z^{-1})y(t) = z^{-d}\frac{B(z^{-1})}{F(z^{-1})}u(t) + \frac{C(z^{-1})}{D(z^{-1})}e(t) \tag{3.12}$$

where $A(\cdot)$, $B(\cdot)$, $C(\cdot)$, $D(\cdot)$, and $F(\cdot)$ are polynomials in the delay operator.

The identification procedure is aimed at determining a good estimate, according to certain criteria, of the two transfer functions introduced in Equation 3.11. This can be done according to the model's ability to produce one-step-ahead predictions with a low variance error. It can be verified that the minimum variance one-step-ahead predictors is (Ljung, 1999)

$$\hat{y}(t \mid t-1) = H^{-1}(z^{-1})G^{-1}(z^{-1})u(t) + \left(1 - H^{-1}(z^{-1})\right)y(t) \tag{3.13}$$

Different families of models, *i.e.* model structures, can be defined by imposing the structure of the transfer functions $G(\cdot)$ and $H(\cdot)$. A model of a given family is determined by identifying the parameters of the transfer functions. The role of the model parameters is clear if the one-step-ahead predictor is rewritten in the following regressor form:

$$\hat{y}(t \mid t-1, \theta) = \varphi(t)\theta \tag{3.14}$$

where

θ is the parameter vector;
φ is the regression vector, which contains past samples of system inputs and outputs and/or residuals, depending on the chosen model structure.

Common dynamic models used in soft sensor applications are now described in the following, by giving the corresponding regression vectors.

The MA model structure is characterized by the following regression vector:

$$\varphi(t) = [u(t-d)...u(t-d-m)]^T \tag{3.15}$$

where d and m indicate the delay of the samples.

This corresponds to

$$\begin{aligned} G(z^{-1},\theta) &= z^{-d}\ B(z^{-1}) \\ H(z^{-1},\theta) &= 1 \end{aligned} \tag{3.16}$$

MA models are characterized by having all the poles at zero; this corresponds to a FIR and can be an adequate model structure for modeling very fast systems.

The regressor for an ARX model is

$$\varphi(t) = [y(t-1),...,y(t-n),u(t-d)...u(t-d-m)]^T \tag{3.17}$$

This corresponds to

$$\begin{aligned} G(z^{-1},\theta) &= z^{-d}\,\frac{B(z^{-1})}{A(z^{-1})} \\ H(z^{-1},\theta) &= \frac{1}{A(z^{-1})} \end{aligned} \tag{3.18}$$

The regressor structure in Equation 3.17 implies that now the $G(\cdot)$ input–output transfer function is not forced to have all the poles at zero. Also the impulse response this time is not extinguished in a finite time span. For this reason, Equation 3.17 is said to correspond to an infinite impulse response (IIR) filter.

Though ARX models can require a smaller number of parameters for accurate modeling of real systems than an MA model, Equation 3.17 clearly shows the necessity to use regressors of the system output. This implies that when data about past output samples are not available, the regressor structure of an ARX model cannot be applied, unless past values of system output are replaced with their estimated values.

When the designer wants to avoid such a choice, an FIR structure is the only suitable solution, with a corresponding growth of model parameters or degradation of model accuracy.

Finally an Auto-Regressive Moving Average, with eXternal input (ARMAX) model is characterized by the regressor:

$$\varphi(t,\theta) = [y(t-1),..., y(t-n), u(t-d),..., u(t-d-m), \\ \varepsilon(t, \theta), ..., \varepsilon(t-k, \theta)]^T \tag{3.19}$$

Note that now the regressor vector depends on the parameter vector θ, making the identification procedure more complex. Equation 3.19 corresponds to:

$$G(z^{-1},\theta) = z^{-d} \frac{B(z^{-1})}{A(z^{-1})} \\ H(z^{-1},\theta) = \frac{C(z^{-1})}{A(z^{-1})} \tag{3.20}$$

The linear structures introduced above can be extended to nonlinear counterparts, that later on in the book will be implemented mainly by using nonlinear neural models based on MLPs. This choice is motivated by the well known approximation capabilities of MLPs with one hidden layer (Haykin, 1999). Nevertheless, case studies will be reported where the complexity of the problems led to a different structure choice.

In particular, if Equation 3.14 is considered, its nonlinear extension is named a NMA model. A block scheme of such a model is shown in Figure 3.4.

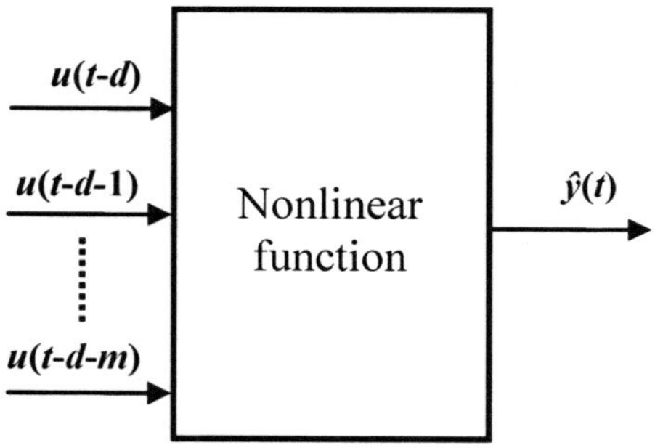

Figure 3.4. Scheme of a NMA model

In the same way, an ARX model can be extended to a NARX model, in accordance with the scheme shown in Figure 3.5.

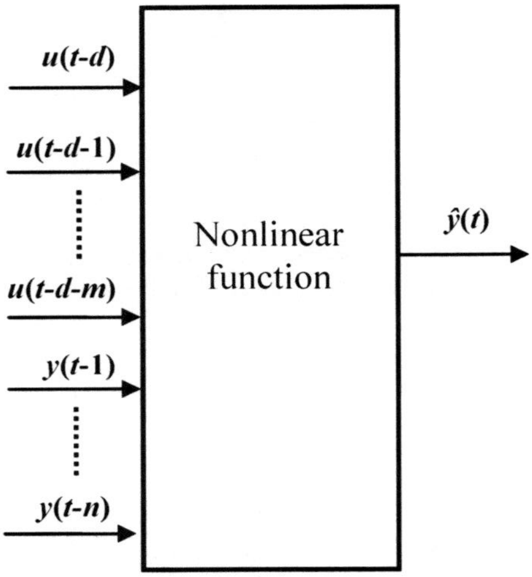

Figure 3.5. Scheme of a NARX model

Finally, the scheme of a Nonlinear ARMAX model (NARMAX) is obtained as an extension of the ARMAX structure, as shown in Figure 3.6.

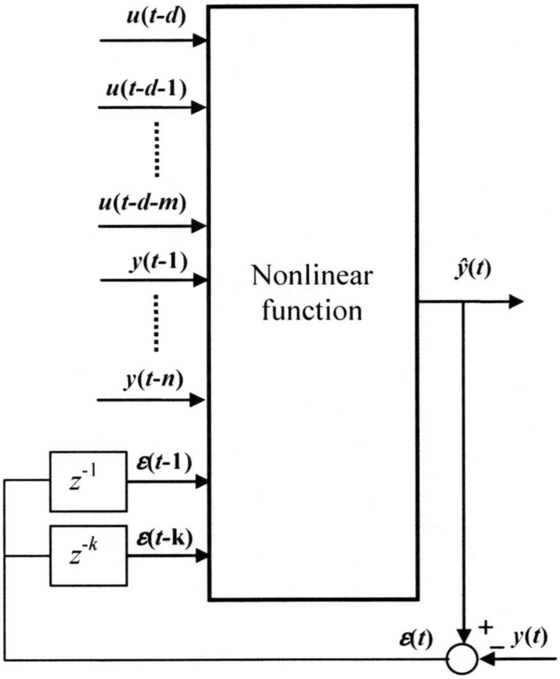

Figure 3.6. Scheme of a NARMAX model

Once the model structure has been selected on the basis of the soft sensor objective and available information, the influential variables and their regressors need to be determined. This is a very complex task and again any physical insight and/or expert knowledge about the process can greatly help effort reduction and a meaningful solution.

Though, as a matter of simplification, these two topics are introduced as separate aspects of the identification process, in practical applications they are tightly intertwined so that, generally, they are solved simultaneously. This means that a large set of variables, including independent variables and a number of their regressors, should be considered as candidate inputs for the model.

A starting list of independent variables and their regressors can be hypothesized by using expert suggestions and any knowledge about system physics. A great help in this phase is when possible a set of experiments designed to gain information about time delays, time constants, and input/output dependence.

Later in this section, a number of methods that allow us to extract a subset of relevant model inputs from the initial set of candidates will be briefly described. In particular, the methods used in the case studies reported in the book will be described, while a number of other approaches can be found in Ljung (1999).

The following approaches will be dealt with (Warne *et al.*, 2004; He and Asada , 1993):

- correlation analysis and scatter plots;
- partial correlation analysis;
- Mallows' statistics;
- Lipschitz quotients;
- PCA and PLS approaches.

Correlation analysis is performed by computing the estimated normalized correlation function between each candidate independent variable (regardless of the delay) and the system output. The magnitude of any peak in the cross-correlation function gives information about the relevance of the input variable (as regards linear dependence), while its position gives information about the correct regressor to be considered in the model. The following MA model is an example of this:

$$y(t) = 0.7\, x(t-10) - 0.3\, x(t-20) \tag{3.21}$$

and the model output has been computed when the input signal is a white noise with a normal distribution. In Figure 3.7, the estimated normalized cross-correlation function obtained in this case is shown. The presence of two large peaks can be observed in Figure 3.7. Also, the position of the peaks is in accordance with the lags introduced in the model 3.21, while their values depend on the coefficients of the model.

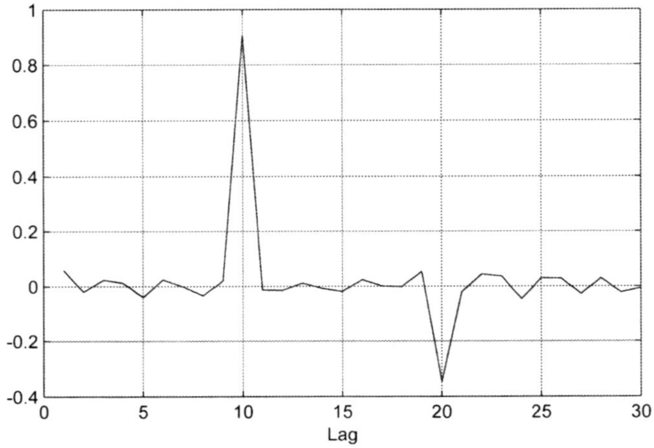

Figure 3.7. Cross-correlation function for the model 3.21, when a white noise with Gaussian distribution is used as input signal

A slightly different kind of analysis is represented by scatter plots. In this case, each candidate variable is plotted against the system output to search for any structure. A straight line will give evidence of a linear input–output correlation, while a curve will suggest a nonlinear relationship. Though the scatter plot method allows us to find also nonlinear input–output correlations, the search for input regressors requires subsequent delays to be plotted: if an input–output correlation exists between the considered input and the system output, the plot that shows a clear structure will give the correct regressor. A slight modification of example 3.21 is considered here, namely the following NMA:

$$y(t) = 0.7\, x(t-10)^2 - 0.3\, x(t-20) \tag{3.22}$$

and again a Gaussian white noise is considered as an input signal for this model. The scatter plot reporting $y(t)$ *vs.* $x(t)$ is reported in Figure 3.8.

It will be observed that no structure can be recognized in the scatter plot. However, this conclusion changes dramatically when the correct lag is considered. In fact, from Equation 3.22 it is evident that a strong correlation between model input and output can be found if input data are delayed by 10 samples. In Figure 3.9, the scatter plot for model 3.22 is reported in the case when $y(t)$ is plotted versus $x(t–10)$.

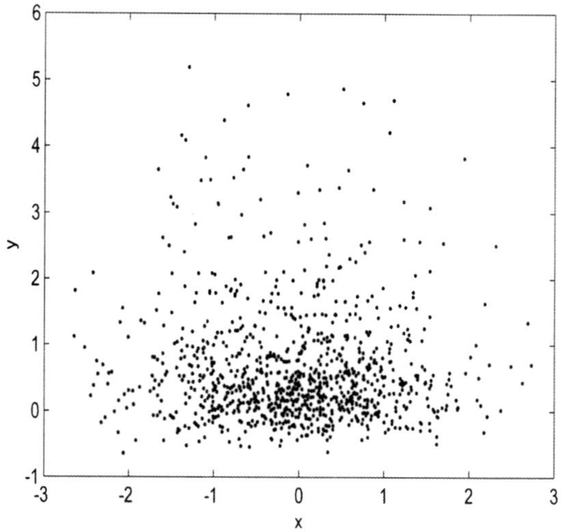

Figure 3.8. *x*–*y* scatter plot for the NMA model 3.22, when a white noise with Gaussian distribution is used as input signal

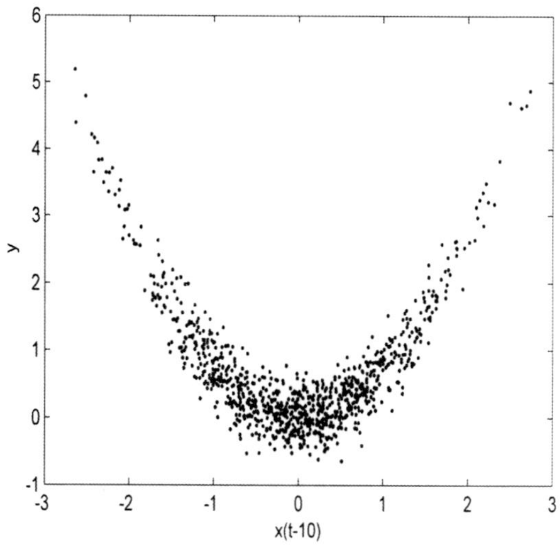

Figure 3.9. $y(t)$ versus $x(t-10)$ scatter plot for the NMA model 3.22, when a white noise with Gaussian distribution is used as input signal

It is evident that in this second case the scatter plot displays a clear structure. Moreover, the points in the scatter plot are approximately parabolic in shape. The dispersion of the points in Figure 3.9 is the effect of the dependence of $y(t)$ from $x(t-20)$. In fact, neither of the methods mentioned so far can filter out the effect of other inputs on the system output. Such an undesired effect can even hide the correlation being sought. Hence, the partial correlation analysis can be used.

The partial correlation method is designed to identify a linear relation between two variables x_i and x_j (or y) "adjusting for" the effects of the remaining variables in the set of candidates. The partial correlation coefficient between the two variables is computed as

$$rp_{ij} = \frac{-C_{ij}}{\sqrt{C_{ii}C_{jj}}} \tag{3.23}$$

where C_{ij} represents the ijth entry of the inverse of the simple correlation matrix.

It is worth noting that this method also gives information about linear correlation, while nonlinear input–output dependence could remain undetected.

The method proposed by Mallow (Warne *et al.*, 2004) is based on evaluation of the standardized total squared error as a criterion for the selection of the model structure. An estimate of model complexity is obtained by using Mallows' C_p statistic, defined as

$$C_p = \frac{R_p}{R} - (n - 2p) \tag{3.24}$$

where:

- n is the total number of model input candidates;
- p is the number of selected variables, extracted from the set of candidates;
- R is the residual variance of the model with the n variables (full model);
- R_p is the residual sum of squares for the reduced model with p inputs.

Good models typically have the (p,C_p) coordinate close to the 45° line on a C_p *vs.* p plot.

An advantage of this method is that it can be used both for linear and nonlinear models.

In Equation 3.24, p is not defined *a priori* and different values need to be analyzed, starting from a minimum value up to n. This is a typical combinatorial problem that can become unmanageable when n is large. It can help in such a case at least to fix p and search for a sub-optimal solution.

The problem of computational load becomes even more severe when the model structure is nonlinear, as in the case of neural network based models. In fact, in this case iterative identification techniques are generally used and the time required for the identification of each model can become considerably longer than the time spent to determine a linear model. A clear example of this drawback is reported in Chapter 5.

Another working method for nonlinear systems is proposed in He and Asada (1993) and is based on the assumption that the system can be represented accurately by a function that is smooth in the regressors.

The method is based on the computation of the so-called Lipschitz quotient for all combinations of input–output pairs and is designed for NARX models:

$$q_{ij} = \frac{|y(t_i) - y(t_j)|}{|\varphi(t_i) - \varphi(t_j)|}, \qquad i \neq j, \ i, j = 1, 2, ..., N \tag{3.25}$$

where N is the number of available samples.

The procedure proposed by He and Asada (1993) is the following:

1. for a given choice of selected model inputs (*i.e.* of model inputs and their regressors) determine the Lipschitz quotients;
2. select the p largest quotients, where $p=0.01\ N \sim 0.02\ N$;
3. evaluate the criterion:

$$\overline{q}^{(n)} = \left(\prod_{k=1}^{p} \sqrt{n} q^{(n)}(k) \right)^{\frac{1}{p}} \tag{3.26}$$

   where n is the number of model inputs;
4. repeat computation 3.26 for a number of different lag structures;
5. plot the criterion as a function of lag space and select the optimal number of regressors as the knee point of the curve.

The procedure described is very time consuming expecially for large sets of candidate inputs, and/or large data sets, though a model need not be identified for each trial.

Once the model structure has been determined, the model parameters need to be identified on the basis of the selected measurement data and a suitable criterion. Minimization of the mean square error (MSE) between the actual system output and its estimation is widely adopted. This topic has been investigated in depth in the literature both for linear and nonlinear models.

In the case of linear models, least mean squares (LMS) methods are generally used, while for nonlinear models the strategy adopted strictly depends on the nature of the nonlinear function selected. In particular, if the nonlinearity is approximated by using MLPs, the identification procedure consists of a suitable learning algorithm.

Though this step of the model identification procedure will not be further addressed in the book, the reader can refer to the widely available literature on optimization algorithms.

PCA and PLS are very powerful tools widely used in soft sensor design. They are applied in a number of steps of a typical design procedure ranging from outlier detection to model identification (Levine, 1996). A number of applications will be described in the book, and a general introduction to theoretical foundations of these techniques will be given here.

PCA is used to rotate original data in such a way to transform correlated variables into a set of uncorrelated ones, which are ordered by decreasing variability. The transformed variables are linear combinations of original ones and the last ones can, eventually, be removed with minimum loss of information.

The first step of PCA consists in data centering and normalization. Scaling data is used to avoid data with large values shading information conveyed by variables represented with smaller numbers. Scaling is generally performed by multiplying each variable by the inverse of its estimated variable standard deviation. Normalized data are therefore organized in a matrix X and the covariance matrix is computed:

$$COV_x = X^T X \tag{3.27}$$

Eigenvectors p_i of this matrix are called loadings and identify the directions where the majority of data variability occurs, while the corresponding eigenvalues give the amount of variability associated with each direction.

The application of PCA to original data produces a new set T of variables, called scores. The relationship between the original data and the corresponding scores is

$$X = T\,P^T \tag{3.28}$$

where P is the matrix containing the eigenvectors of the covariance matrix 3.27.

Equation 3.28 holds true only in the case when the whole set of eigenvectors is used. If a reduced set is of interest, scores can only approximate original data:

$$X = T\,P^T + E_x \tag{3.29}$$

where E_x is the reconstruction error. Interestingly, based on the order given to the transformed matrix, by considering the first scores in the reduced models, most of the variability of the original variables is retained.

As an example, the scatter plot of a synthetic data set, consisting of a dimension two matrix, is now considered. Probably, this is not the most suitable example, because PCA is more useful when multivariate problems with a large number of variables are considered. In fact, in these cases the possibility to reduce the size of the problem is greatly welcomed. Nevertheless, considering a dimension two problem allows us to give a graphic representation of the data and of the effect of PCA application.

The scatter plot of the data considered is shown in Figure 3.10. Data reported on the vertical axis were obtained by multiplying data on the horizontal data axis by a constant value and then adding a suitable amount of noise. This produces a significant correlation among data. The directions of the loadings are also reported, with the use of solid lines.

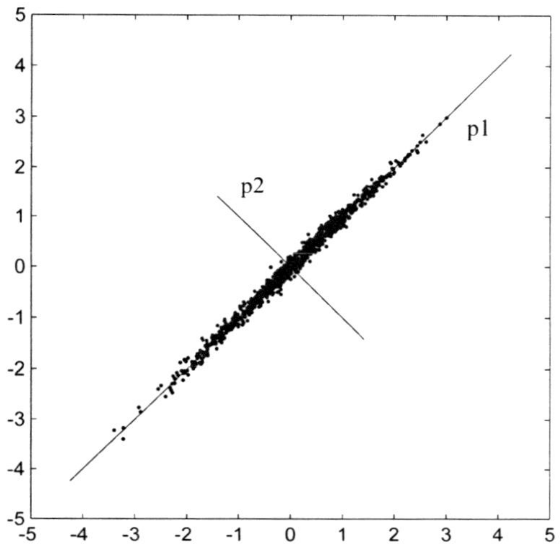

Figure 3.10. Scatter plot of two correlated synthetic vectors. Scores are indicated with two solid lines.

The scatter plot of corresponding latent variables obtained after application of PCA to the original data is reported in Figure 3.11.

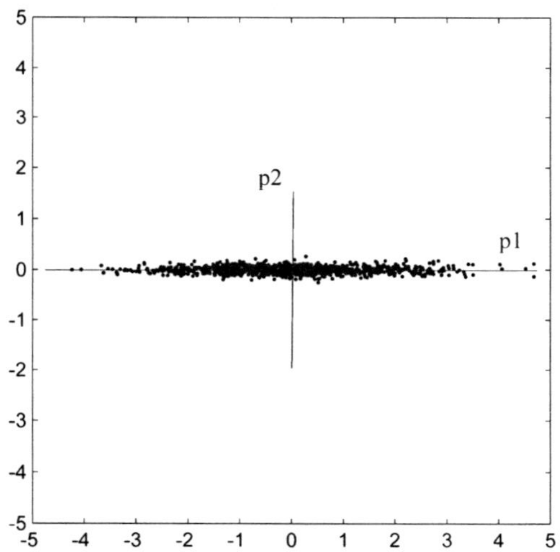

Figure 3.11. Scatter plot of latent variables obtained by applying PCA to data shown in Figure 3.10

Interested readers can find details about PCA algorithms in the cited bibliography.

In contrast to PCA, PLS considers information about both independent and dependent variables, in order to determine a linear model, by using sets X and Y of measurements, both affected by noise.

In this case, both the X and Y sets are transformed into scores and loadings, so that two new sets (T and U) are obtained.

To obtain the PLS of original data, X is modeled by using the first principal component:

$$X = t_1 p_1^T + E_{x,1} \tag{3.30}$$

Similarly, Y is modeled as

$$Y = u_1 q_1^T + F_{y,1} \tag{3.31}$$

Residuals matrices $E_{x,1}$ and $F_{y,1}$ are therefore modeled using again Equations 3.30 and 3.31 to find t_2, p_2, u_2, and q_2. The procedure can be further repeated, until no useful information is left in the residuals.

The procedure described returns the scores t and u and the loadings p and q of X and Y, respectively. The scores of input and output variables are related by the following internal linear model:

$$u_i = b_i t_i + r_i \tag{3.32}$$

Finally, it is worth noting that PLS returns a linear model for the system. Nevertheless, extensions to nonlinear PLS (NPLS), based both on polynomial functions and on neural networks, have been proposed in the literature, (Baffi, Martin and Morris, 1999).

Case studies where NPLS have been applied to real industrial applications will be described elsewhere in the book.

3.5 Model Validation

The final step towards the identification of a model is its validation. Validation is the phase required to verify whether the model is able to adequately represent the underlying system.

A general rule to be followed when approaching this phase of soft sensor design is that data used for model validation should be different from those used for model identification. In fact, a model could have satisfactory behavior with the learning data set and work very poorly when processing a new data set. This precaution is useful for investigating overfitting phenomena. Overfitting should anyway be taken into account during the model identification procedure by using adequate strategies and/or algorithms (*e.g.* early stopping), especially when the amount of available data is small compared to the model complexity.

The task of model validation is a complex one and it becomes even harder in the case of validation of a nonlinear model, which is often the case for a soft sensor designer. Any valid help is greatly welcome in this phase and, generally, expert designers use a mixture of the different techniques available for data analysis.

The criteria to be used during the validation step can be divided into two groups. In the first one, general techniques devoted to an analysis of the model residual properties can be found, while the second one groups criteria that are suggested by the application the soft sensor is designed for.

Here techniques from the first group will be described. Examples of the second group are strictly linked to the application and will be given later on in the book, when we proceed to the critical analysis of the performance of models obtained during the solution of corresponding case studies.

Techniques for model validation are aimed both at analyzing statistical model residual characteristics and at searching for any undesired correlation between model residual and present and/or delayed samples of model inputs and outputs. Notwithstanding this general aspect, validation of nonlinear models requires an even deeper analysis than linear models.

In particular, graphical approaches can be very powerful tools for model validation, especially if adequate graphics are presented in a number of plots that allow the user to extract information from direct data inspection.

The very first, and often the most important, graphical analysis consists in the visual inspection of time plots of both recorded data and their estimation, on a large and significant set of data. This can help in having an immediate feeling for the capability of the model to reproduce significant system dynamics. Such a test needs to be performed both on learning and test data. Moreover, attention must be paid to select data trends that are relevant for the intended soft sensor application.

Graphical techniques can also help to gain evidence that residuals between the measured output and the estimated output contain no information about past residuals or about system dynamics, as in the case of correlation graphs, and that the residuals have specific distribution, location and variation (Chen, Billings and Grant, 1990; Billings, Jamaluddin and Chen, 1992; Guidorzi, 2003; Ljung, 1999; NIST/SEMATECH, 2005).

A simple technique that can help to test the above mentioned assumption is known as the 4-plot analysis of model residuals.

In more detail, the 4-plot analysis is based on visual inspection of the following graphs:

- run sequence plot;
- lag plot;
- histogram;
- normal probability plot.

These four graphics can be used to test specific properties of the model residual. In particular, if the run sequence plot, *i.e.* the time plot of the model residual, does not show any drift nor any change in its spread, then the assumptions of a residual with fixed location and variation hold true.

The lag plot is used to search for any dependence of the model residual on its past values. This, in fact, will be reflected in the presence of a structure in the lag plot. On the contrary, the randomness assumption will produce a lag plot without any relevant structure.

The histogram and the normal plot will be used to investigate the normal distribution of the model residual. In fact, if the model identification works properly, the histogram will be approximately zero centered and bell-shaped, while the normal plot will be close to a straight line.

Any discrepancy from the 4-plot ideal aspect is a symptom of divergence from the stated hypothesis for the model residual, so that the designer should go back to the flow diagram reported in Figure 3.1 and search for anything wrong in the design process. Also, each cause of departure from the ideal case produces specific (and different) effects on the 4-plot, so that from the 4-plot analysis it is possible to determine the cause of modeling failure.

In Figure 3.12, an example of the 4-plot for a synthetic random variable is reported. A visual inspection of the plots shows that all the considerations reported above are satisfied.

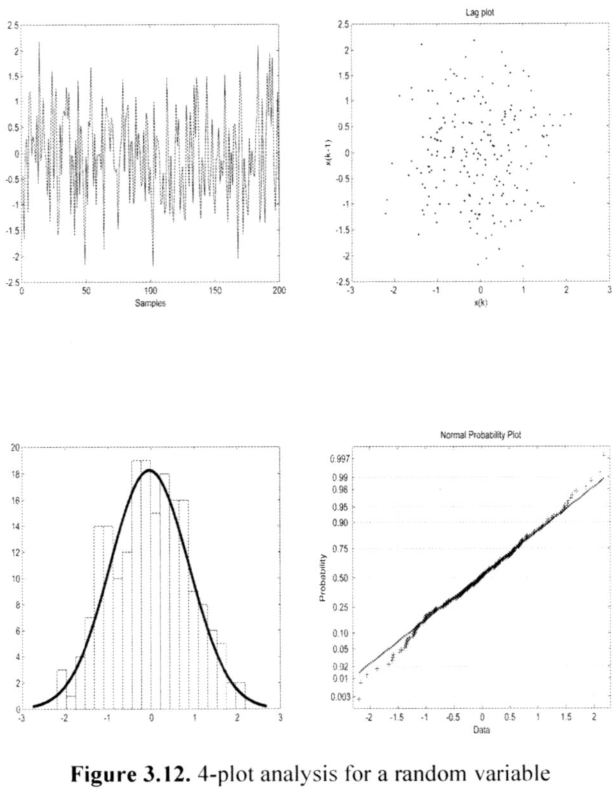

Figure 3.12. 4-plot analysis for a random variable

An example of a 4-plot, obtained for a variable that was not normally distributed, is reported in Figure 3.13. In this case an oscillatory variable was synthesized and a noise signal, with suitable variance, was added. Visual perusal of the graphs shown in Figure 3.13 allows a number of interesting conclusions to be drawn. The run sequence plot shows a low-frequency fluctuation that evinces evidence of variation of residual location. Also a structure is clearly visible in the lag plot. The histogram shows a bimodal (U-shaped) distribution. Finally, the normal probability plot shows evident departures from the straight line, especially on the tails.

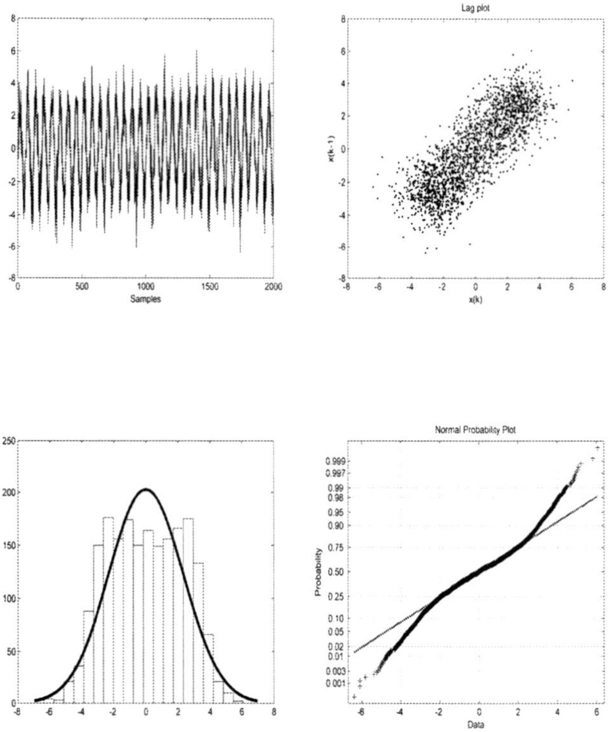

Figure 3.13. 4-plot analysis for a noisy oscillatory variable

In Figure 3.14, results of the case study described in Chapter 5 (Figure 5.37) are reported here as an example of 4-plot analysis of the residuals of a model for a real industrial application.

More 4-plot analyses performed on residuals of identified models for soft sensor implementation will be shown in the following chapters of the book.

A number of further examples of 4-plot analysis can be found in NIST/SEMATECH (2005), where 4-plots are shown along with a description of the conclusions that can be drawn from visual perusal of the graphs.

The procedures that should be followed in the case of linear systems are quite well established and a huge amount of literature exists on this topic (Ljung, 1999).

When dealing with nonlinear systems, the checking of residual properties is more cumbersome; however, a number of tests designed to detect whether the residual is unpredictable from all linear and nonlinear combinations of past inputs and outputs is reported in the literature (Billings, Jamaluddin and Chen, 1992; Billings and Voon, 1991; Mendes and Billings, 2001). These tests were introduced considering the case of analytic nonlinear models; they are generally used for neural network based NARX models. In fact, while a theoretical analysis of this aspect does not exist, experimental evidence of their usefulness in the case of neural models is widely available.

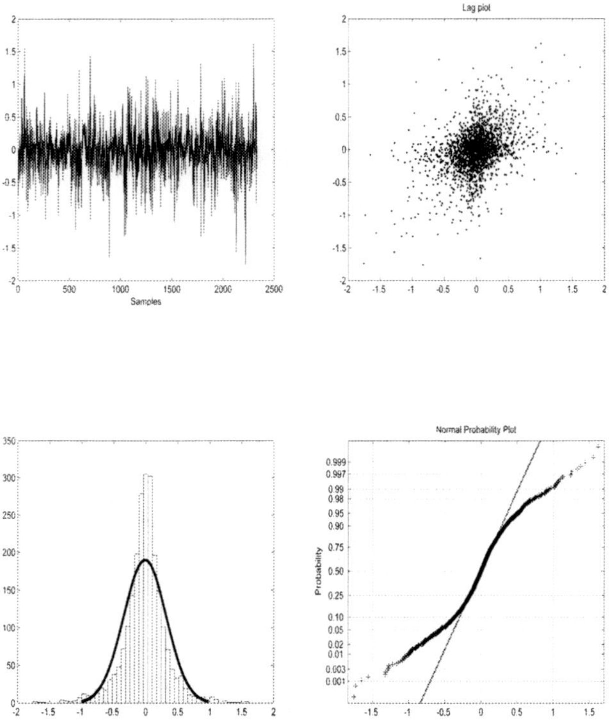

Figure 3.14. 4-plot analysis for the real case study reported in Chapter 5 (Figure 5.37)

In practice, model validation requires that normalized correlation functions between couples of sequences ψ_i and ψ_j be estimated. The sampled correlation function is

$$\hat{\phi}_{\psi_i\psi_j}(\tau)=\frac{\sum_{t=1}^{N-\tau}\psi_i(t)\psi_j(t+\tau)}{\sqrt{\sum_{t=1}^{N}\psi_i^2(t)\sum_{t=1}^{N}\psi_j^2(t)}} \tag{3.33}$$

The following conditions need to be tested, for each input variable:

$$\phi_{\varepsilon\varepsilon}(\tau) = E[\varepsilon(t-\tau)\varepsilon(\tau)] = \delta(\tau) \tag{3.34}$$

$$\phi_{u\varepsilon}(\tau) = E[u(t-\tau)\varepsilon(\tau)] = 0 \quad \forall\tau \tag{3.35}$$

$$\phi_{u^{2'}\varepsilon}(\tau) = E[(u^2(t-\tau) - u_m^2(t))\ \varepsilon(t)] = 0 \quad \forall\tau \tag{3.36}$$

$$\phi_{u^{2'}\varepsilon^2}(\tau) = E[(u^2(t-\tau) - u_m^2(t))\varepsilon^2(t)] = 0 \quad \forall\tau \tag{3.37}$$

$$\phi_{\varepsilon(\varepsilon u)}(\tau) = E[\varepsilon(t)\ \varepsilon(t-1-\tau)\ u(t-1-\tau)] = 0 \quad \tau \geq 0 \tag{3.38}$$

where $E[\cdot]$ is the mathematical expectation, $\varepsilon(t)$ is the model residual, $u(t)$ is the generic input variable, and the subscript m denotes the time average. The normalization adopted in Equation 3.33 guarantees that introduced correlation functions are in the interval [–1, 1].

In practical applications, none of the reported correlation functions are exactly zero for any of the lags considered. Instead, they are traced with a corresponding confidence band. The 95% confidence band is generally used and corresponds to $1.96/\sqrt{N}$, where N is the number of samples considered.

The presence of values of correlation functions lying significantly outside the confidence band for any time lag suggests that it is advisable to consider the corresponding lagged input in the model structure. The occurrence of this undesirable behavior should be checked for the first delay values, while for larger values of the delay the physical meaning of the cross-correlation function can be lost.

If the validation phase fails, because of any of the reported criteria, the soft sensor design procedure should be reconsidered. In particular, changes in any step of the procedure should be taken into account. As examples, identification procedure failure could have been caused by a wrong selection of data used for model identification, incorrect outlier detection, inadequate choice of influential variables, and so forth.

In the following chapters, a number of industrial case studies, based on real data and intended to highlight each step of the identification procedure, are reported.

4

Selecting Data from Plant Database

4.1 Detection of Outliers for a Debutanizer Column: A Comparison of Different Approaches

In this chapter, the detection of outliers, a topic addressed from the theoretical point of view in Chapter 3, is reconsidered by taking a case study. Some of the methods discussed will be applied to data referring to the debutanizer column described in the Appendix.

The purpose of modeling in the case considered was to obtain a very accurate dynamic model able to estimate the system output in real time, avoiding the long delay introduced by the analyzer, which in this case was estimated at about 45 min.

Experts furnished a set of ten input variables which were considered relevant (procedures to be followed to select input variables will be addressed in Chapter 5). Data were acquired in different periods, with a sampling time $T_s = 6$ min; in this chapter, results obtained for a subset recorded in one period will be described. A set of about 2500 data values (normalized so as to avoid important variables whose magnitudes are small from being overshadowed by less important but larger magnitude variables), extracted from the plant historical database was considered. The same data set was processed with all the methods considered for the purposes of comparison.

The following methods were applied:

- 3σ edit rule;
- Jolliffe parameters with PCA;
- Jolliffe parameters with PLS;
- Residual analysis of linear regression.

The methods listed above are those most commonly suggested in the literature, though others can be found. However, their nonlinear counterpart can be considered, in particular nonlinear PLS regression and residual analysis of nonlinear regression can be considered.

The data were preprocessed in order to take into account the presence of missing and invalid data, recognized by the acquisition system and denoted as Out_of_Range.

Whatever the technique used for outlier detection, a post-processing phase is necessary to remove outliers. Two strategies can be adopted depending on the size of the cluster of detected outliers. If an isolated outlier is detected it can be replaced; a common strategy is to use interpolation of adjacent data. When a cluster of consecutive data is detected as outliers, interpolation can be meaningless and a common strategy is to eliminate them and resize the data set. In this case, attention must be paid when searching for dynamical models because these models will use as input, time regressors of the input and/or output variables, and jumps in timescale cannot be introduced. In this case, the right choice may be to maintain the data sets as different files to be fused only after the regressors have been built.

4.1.1 The 3σ Edit Rule

The 3σ edit rule is based on the hypothesis that the variable considered is normally distributed and it is generally applied even though the normality condition is seldom satisfied in practice. Usually this method is made robust by replacing the outlier-sensitive mean and standard deviation estimates with the outlier-resistent median and median absolute deviation from the median (MAD), respectively.

In the case of the debutanizer, computations have been performed with respect to mean and standard deviation estimates because suspected outliers were not sufficiently relevant to modify these parameter estimations.

The 3σ edit rule method has the valuable property of allowing visual inspection of each input and output variable, without projections that could make it hard to understand the physical meaning of the results obtained. Also, comparative inspection of the graphs obtained for the process variables allows one to decide, with the help of a process expert, whether a candidate outlier is due to a particular working condition (anomalies will be seen, in this case at least, in a subset of available graphs), or to sensor/transmission errors. In the former case, of course, the anomalous behavior reflects a working condition of the plant, even if a rare one, and it may or may not be considered an outlier, depending on the modeling task. In fact, taking into account that the data used must reflect all the possible dynamics of the system, such rare events can represent a precious source of information for the process of model identification.

In the following Figures 4.1 to 4.11, time plots of the dependent variable and of corresponding independent variables are shown. Detected outliers are reported in gray, along with the corresponding order number. Each figure also reports the corresponding histogram to check the validity of the normal distribution hypothesis.

As an example, visual inspection of Figure 4.1 shows that the suspected outliers for the process output variable are actually part of higher peaks that could contain valuable information about the plant dynamics. In such conditions, the final decision should be left to the experts. A different condition can be hypothesized for the data reported in Figure 4.3, where detected outliers belong to a cluster of data clearly separated from the general trend of the considered variable.

The reported histograms clearly show that the hypothesis of normal distribution is partially verified.

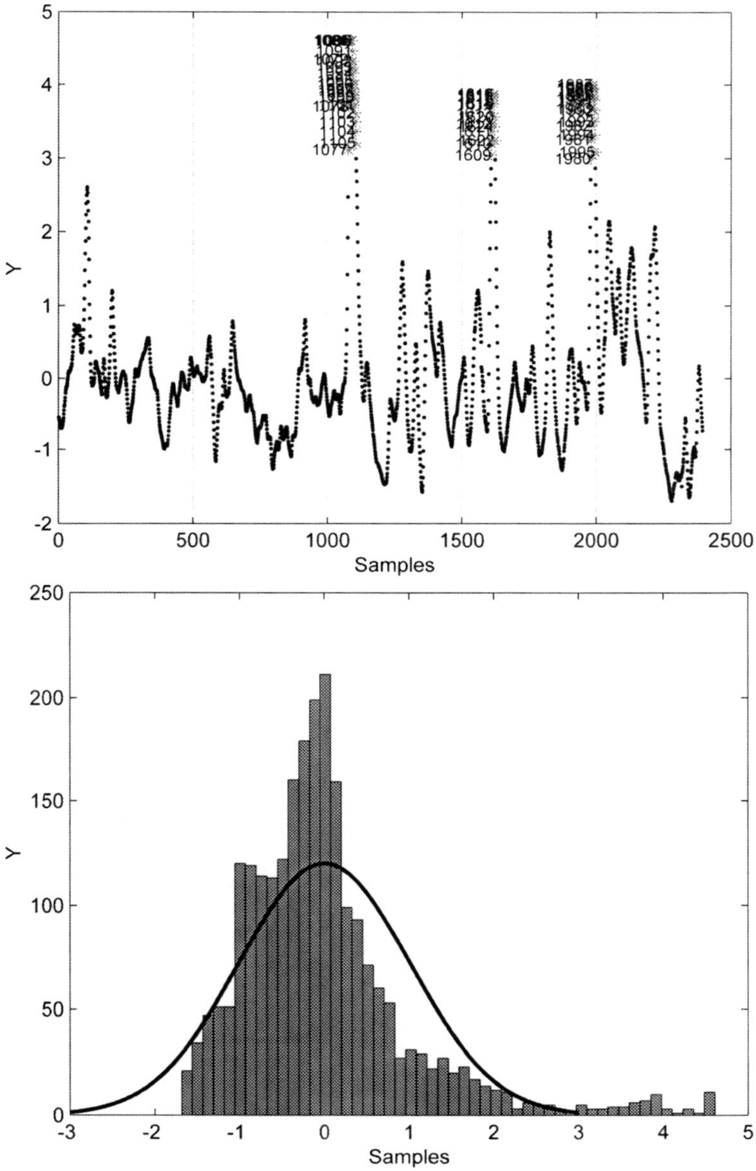

Figure 4.1. Results of the 3σ edit rule for the dependent variable of the debutanizer column

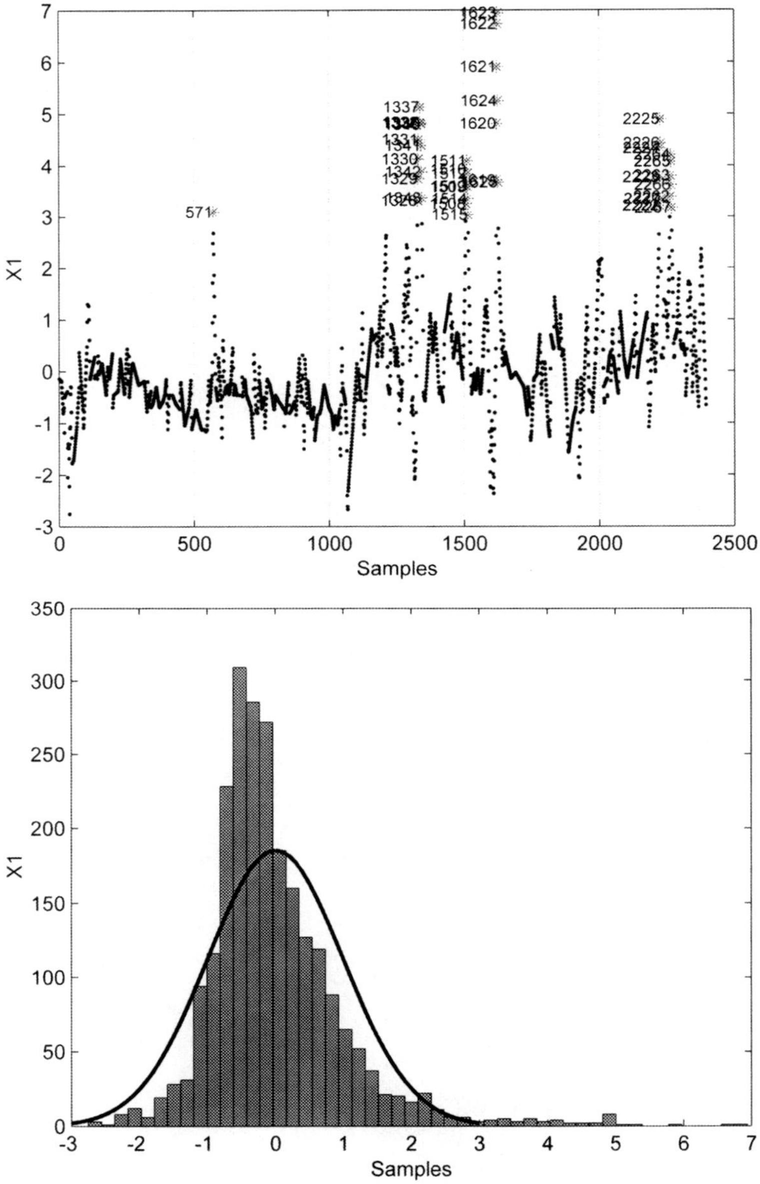

Figure 4.2. Results of the 3σ edit rule for first independent variable of the debutanizer column

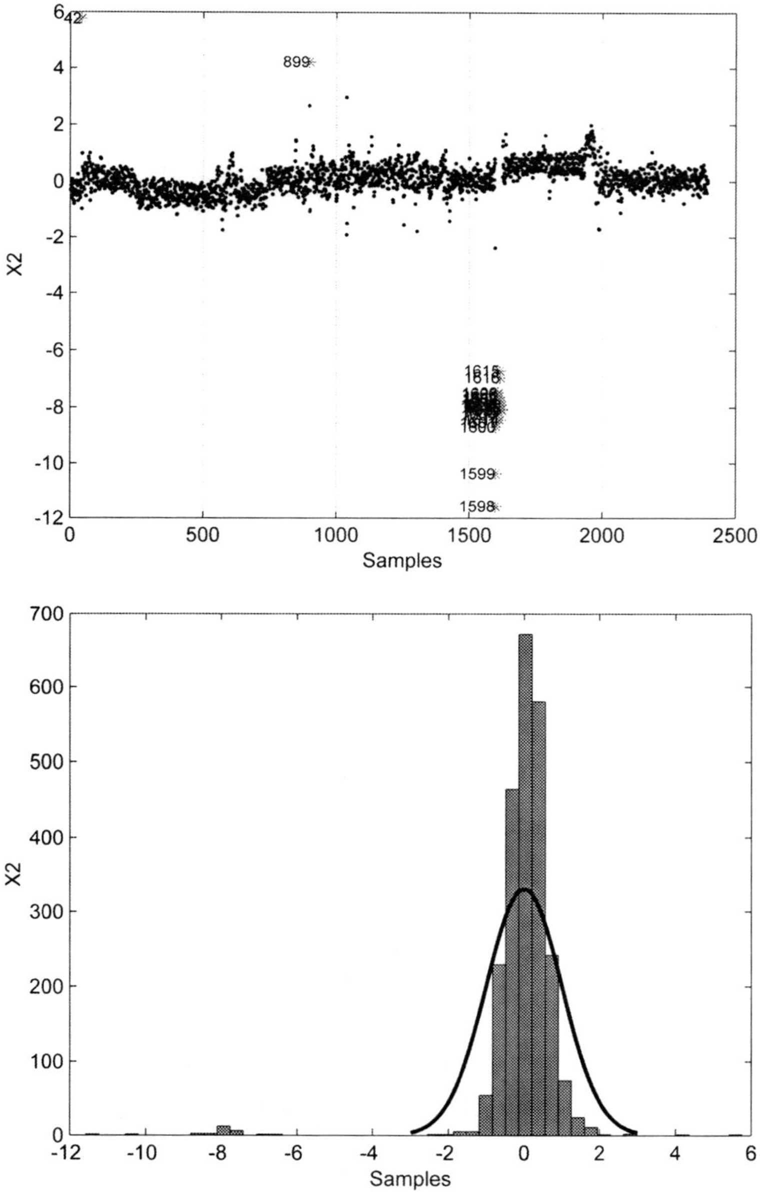

Figure 4.3. Results of the 3σ edit rule for the second independent variable of the debutanizer column

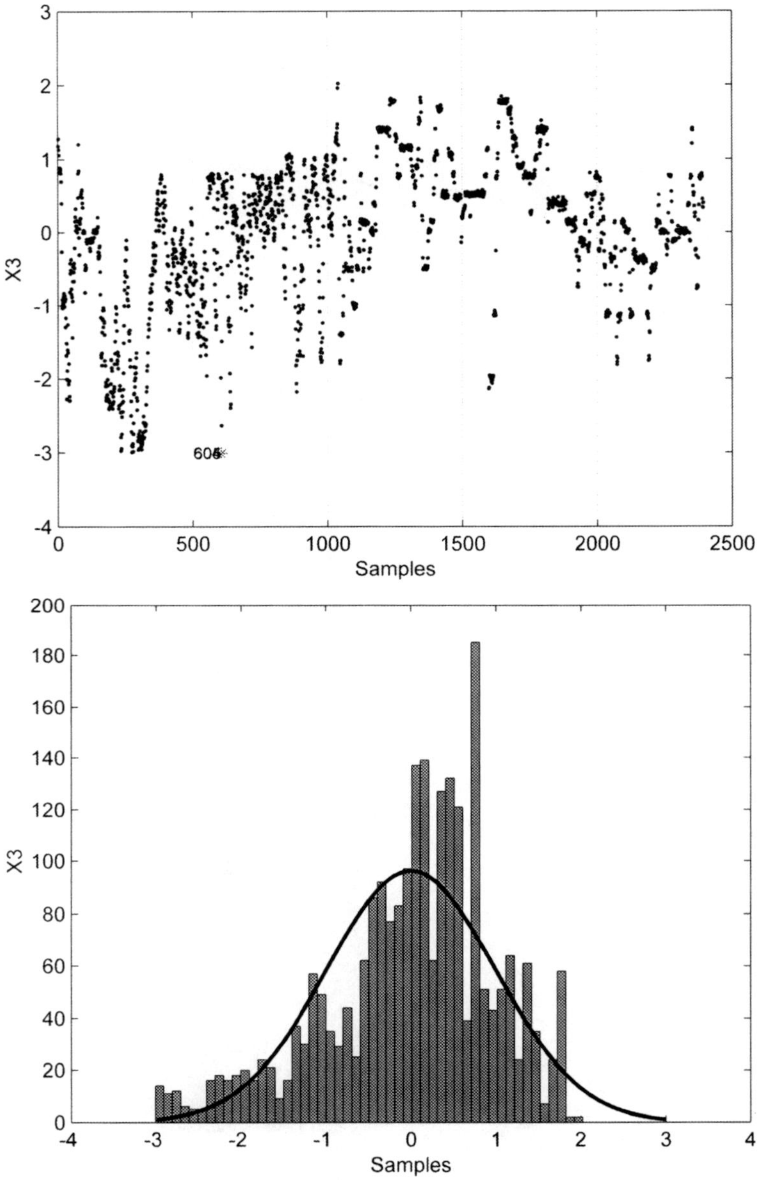

Figure 4.4. Results of the 3σ edit rule for the third independent variable of the debutanizer column

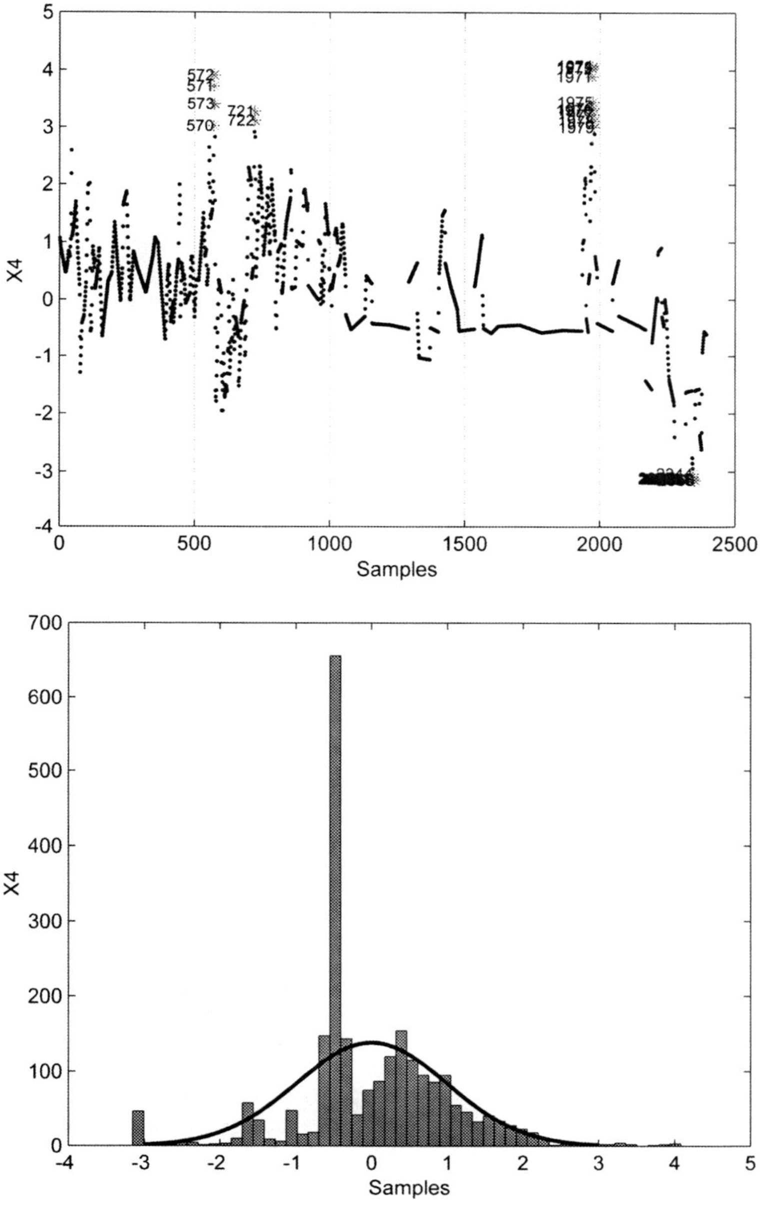

Figure 4.5. Results of the 3σ edit rule for the fourth independent variable of the debutanizer column

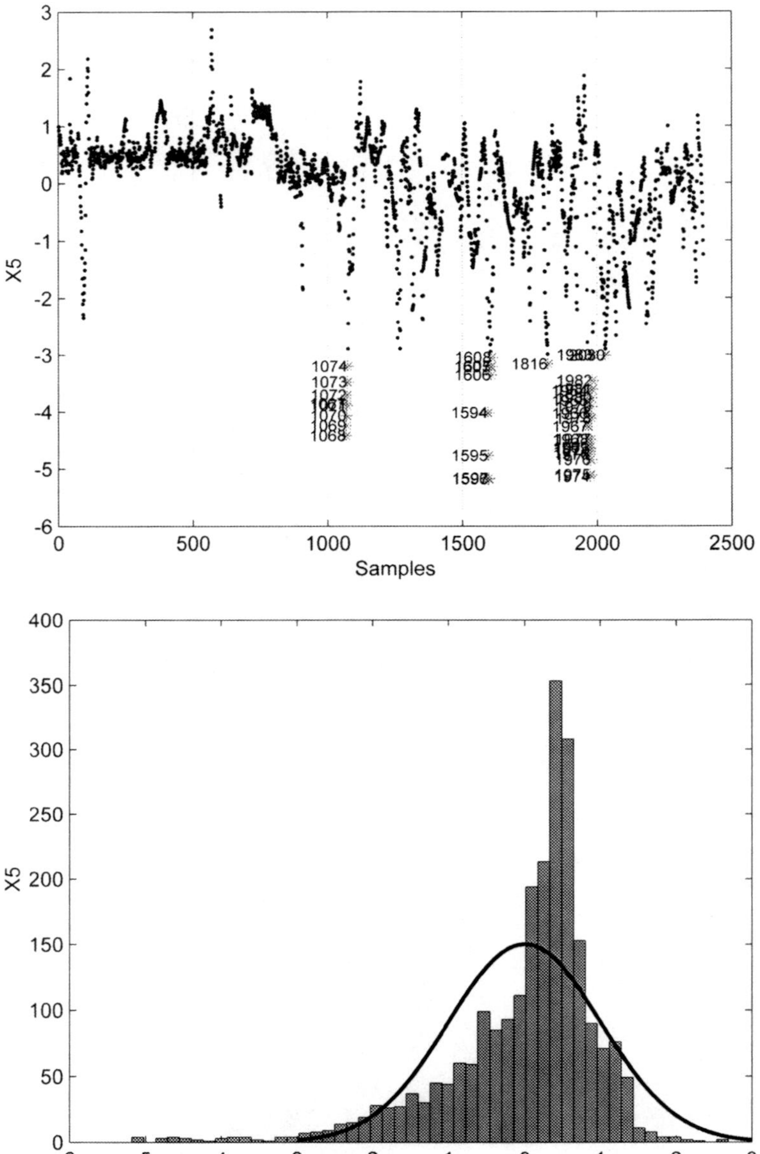

Figure 4.6. Results of the 3σ edit rule for the fifth independent variable of the debutanizer column

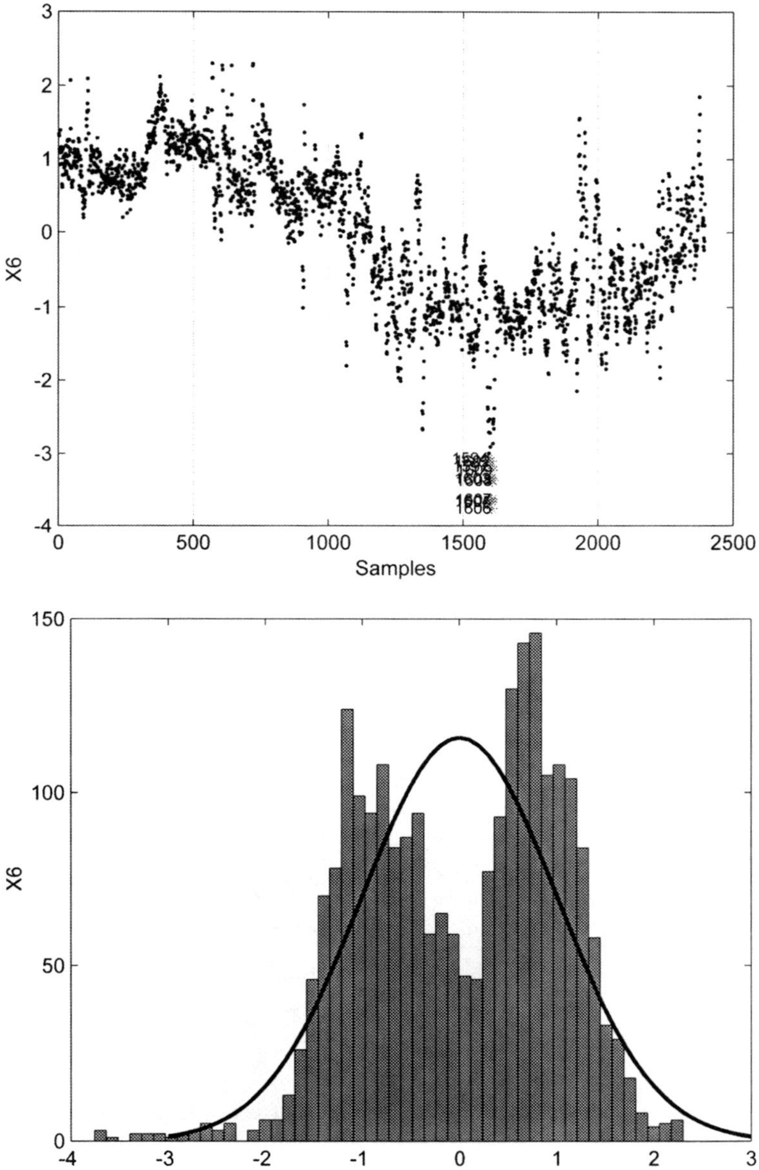

Figure 4.7. Results of the 3σ edit rule for the sixth independent variable of the debutanizer column

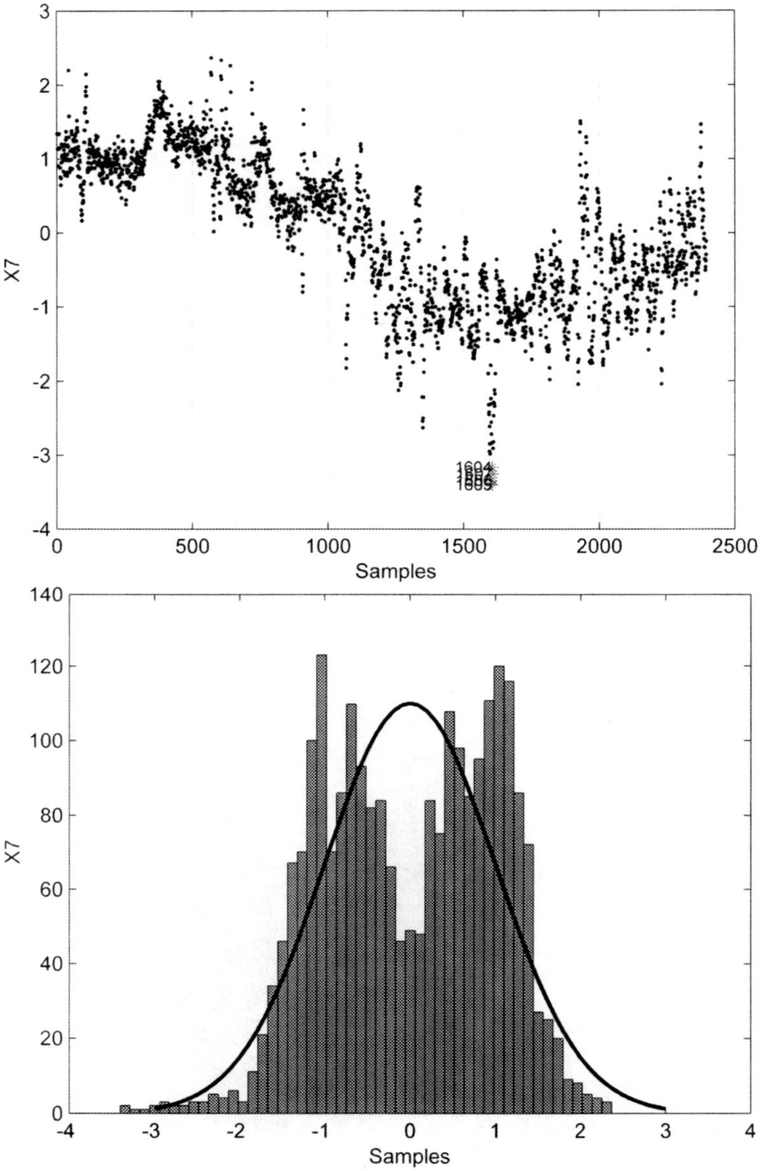

Figure 4.8. Results of the 3σ edit rule for the seventh independent variable of the debutanizer column

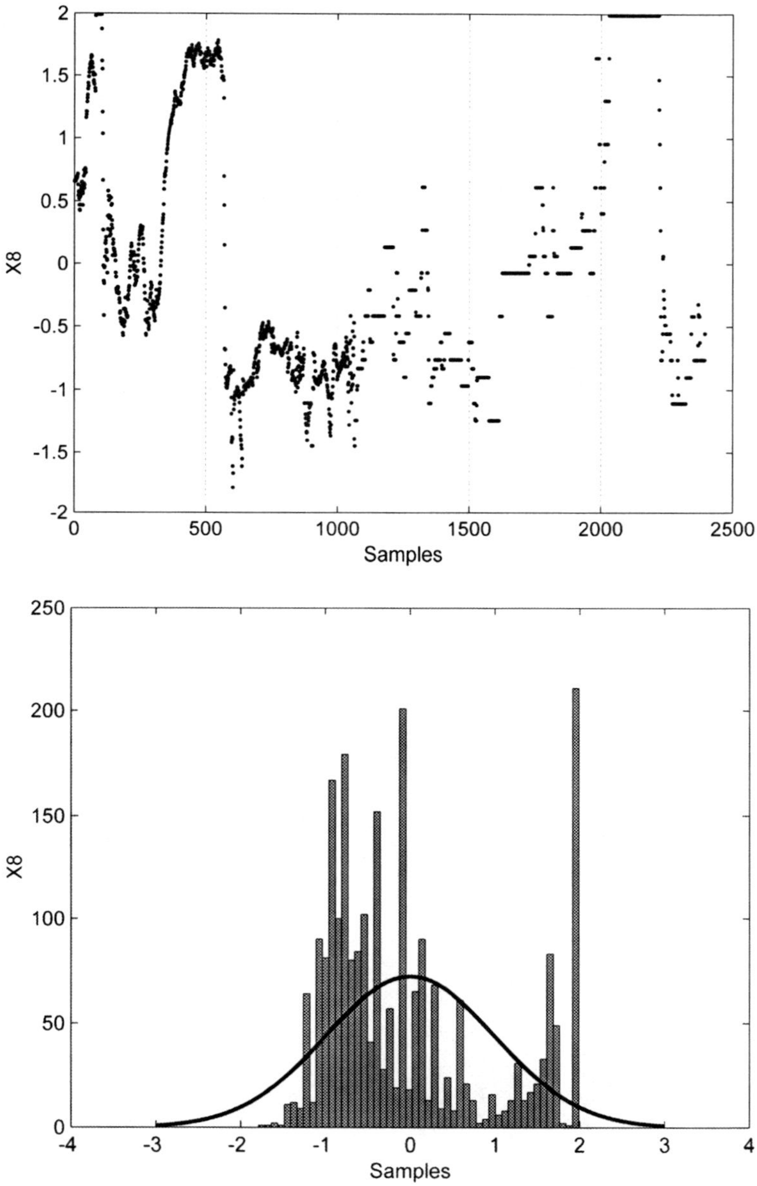

Figure 4.9. Results of the 3σ edit rule for the eighth independent variable of the debutanizer column

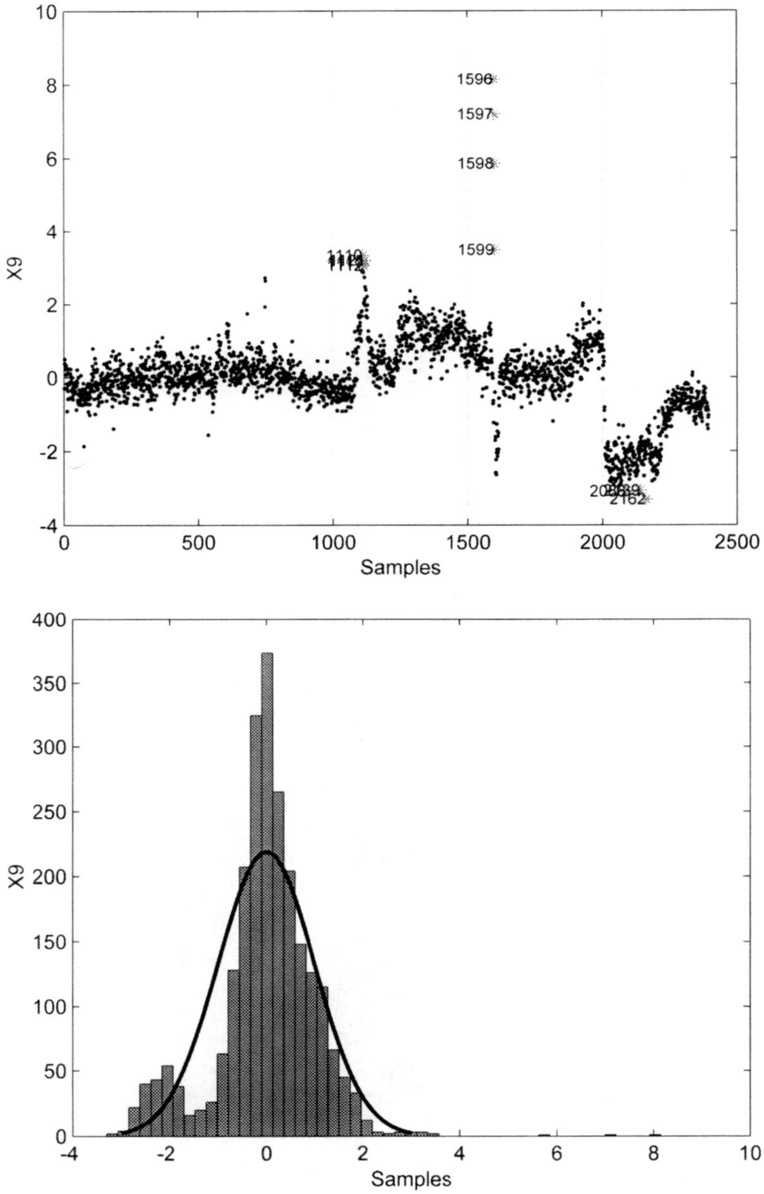

Figure 4.10. Results of the 3σ edit rule for the ninth independent variable of the debutanizer column

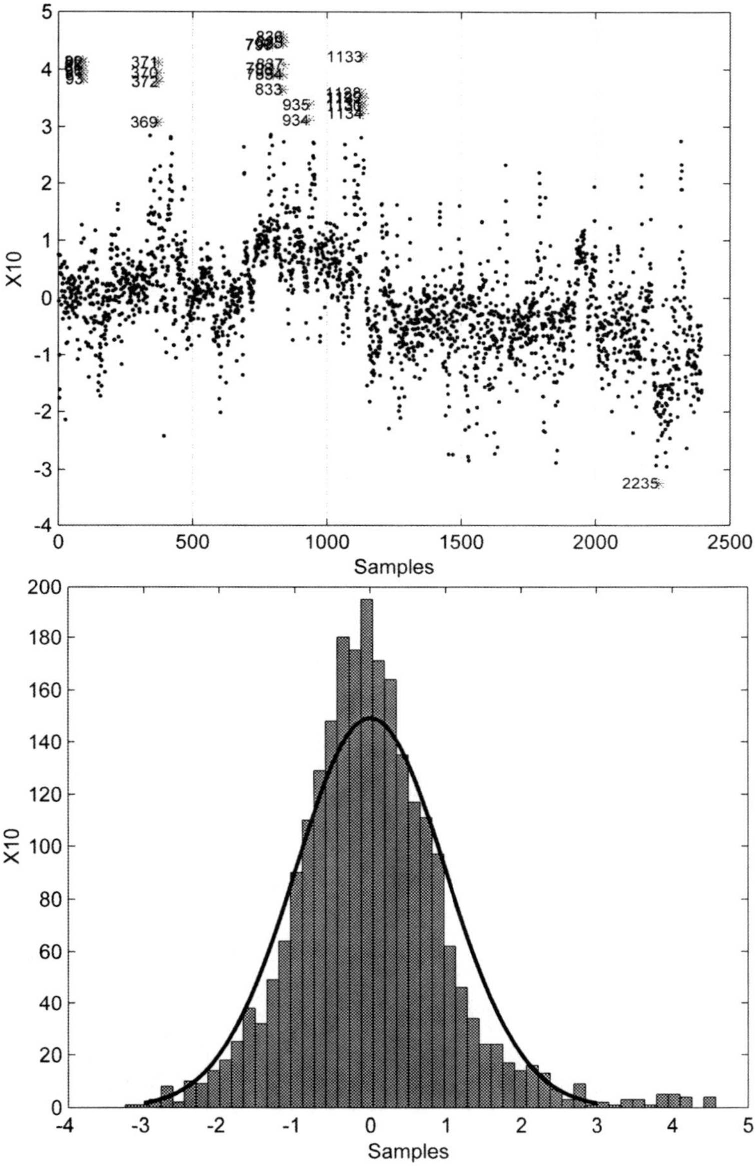

Figure 4.11. Results of the 3σ edit rule for the tenth independent variable of the debutanizer column

4.1.2 Jolliffe Parameters with Principal Component Analysis

As reported in Chapter 3, outliers can be searched for using parameters suggested by Jolliffe, after PCA or PLS decomposition of original data have been performed. In this subsection we describe results obtained by applying PCA decomposition to data referring to the debutanizer column, while results obtained using PLS will be reported in the next subsection. When using Jolliffe parameters, a suitable limit to detect outliers should be defined. We used the generally adopted 3σ limit for each Jolliffe parameter. The results are shown in Figures 4.12, 4.13, and 4.14.

As can be observed from comparison of the three Jolliffe parameters, they mostly agree about detected outliers, notwithstanding that d_1^2 has been designed to penalize more severely components with a low value of variance, with respect to d_2^2, and that statistic d_3^2 is designed to detect observations that inflate the data set variance.

A simpler approach to outlier detection, commonly used in the literature, consists in the direct visual analysis of the graph reporting the first principal component *vs.* the second component. Isolated points should be considered as suspected outliers. Figure 4.15 was obtained for the application considered. In the figure, labeled points refer to samples recognized as outliers by all three Jolliffe parameters.

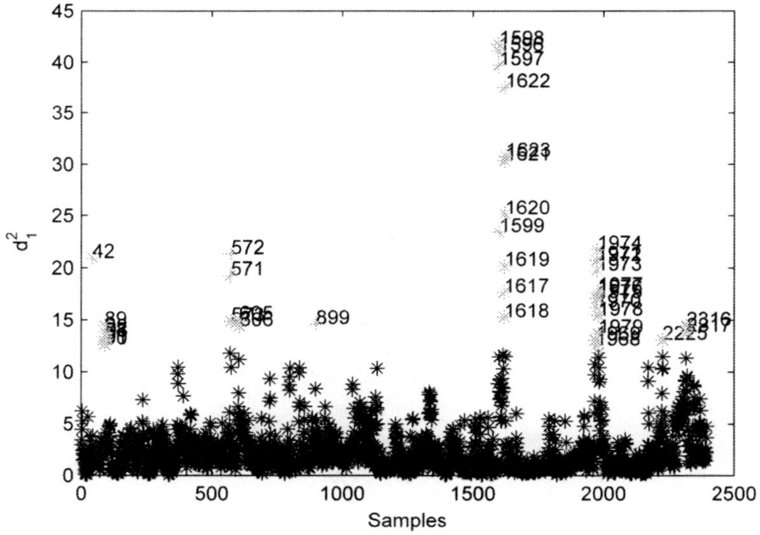

Figure 4.12. The first Jolliffe parameter (d_1^2), computed on the set of data taken from the debutanizer column, transformed using PCA

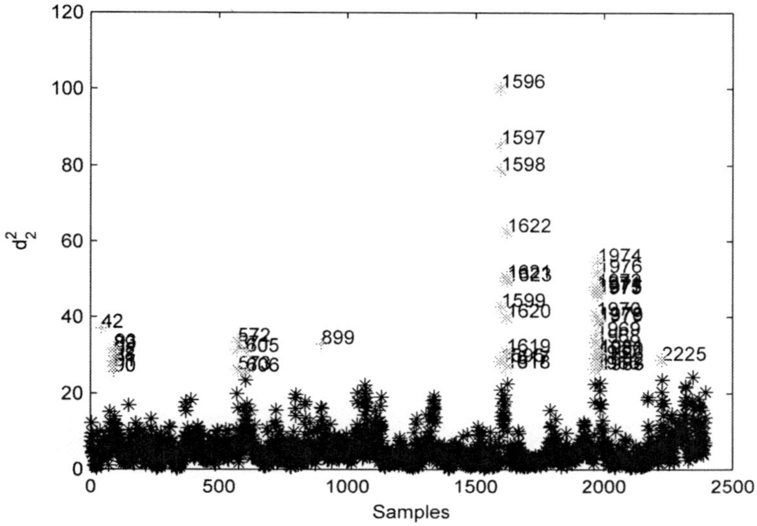

Figure 4.13. The second Jolliffe parameter (d_2^2), computed on the set of data taken from the debutanizer column, transformed using PCA

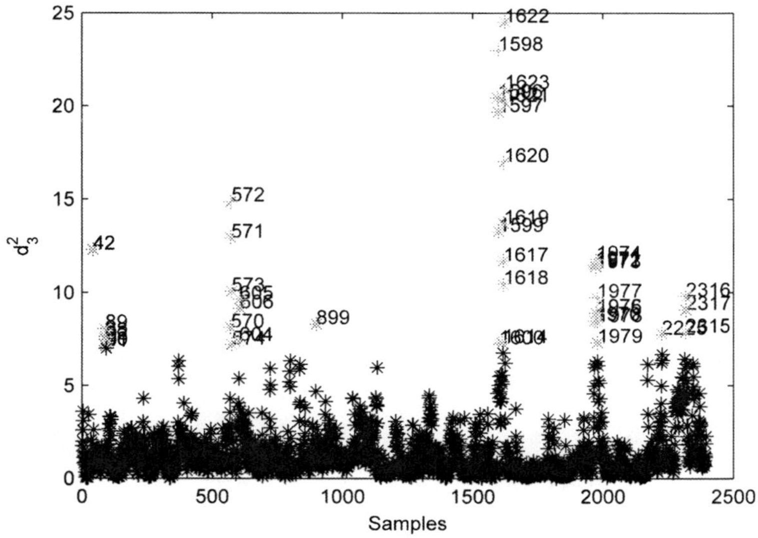

Figure 4.14. The third Jolliffe parameter (d_3^2), computed on the set of data taken from the debutanizer column, transformed using PCA

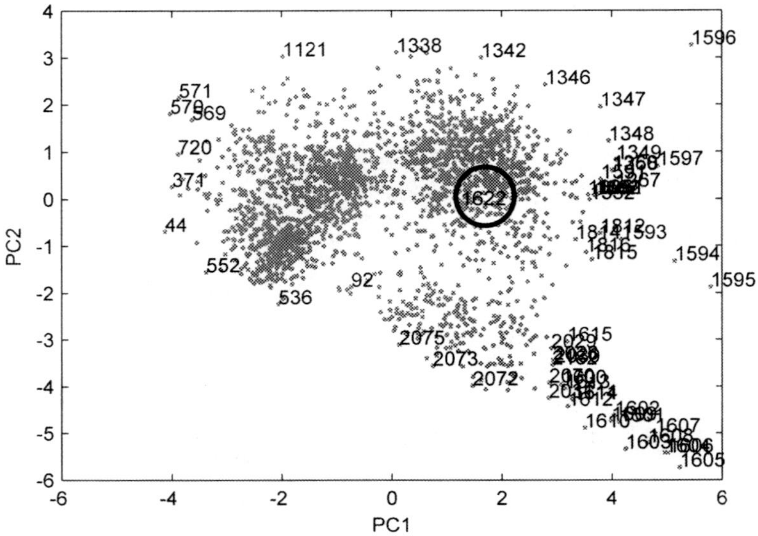

Figure 4.15. Plot of the first principal component (PC1) *vs.* the second principal component (PC2), for the debutanizer column

As can be observed, direct inspection of the PCA plot does not give any automatic suggestion about outlier detection, which is somewhat left to the designer, even if suitable limits can always be defined (Chiang, Perl and Seasholtz, 2003).

In Figure 4.15 the encircled point corresponds to an outlier which was detected by Jolliffe parameters and that is totally lost when using direct PCA plot . This is a case when the methods are not totally equivalent, even when the limits of the PCA component plot are fixed. Readers can also refer to the book site to download data of variables involved and look at data plots to identify the corresponding outliers.

A common drawback of both the Jolliffe parameters and direct PCA methods is that no information about system output is taken into account in the decision process. Last, but not least important, the physical meaning of the input/output variable is lost and this makes it harder to involve process experts in the detection step, unless back projection to the original data is performed.

An in-depth comparative analysis of the various methods will be given at the end of this chapter.

4.1.3 Jolliffe Parameters with Projection to Latent Structures

Correlation between input and output variables can be taken into account by using PLS, instead of PCA. Among other techiques, Jolliffe parameters can be used also in this case to make it easier to detect outliers after PLS projection of original data. The results of applying PLS to the Jolliffe method are shown in Figures 4.16, 4.17, and 4.18.

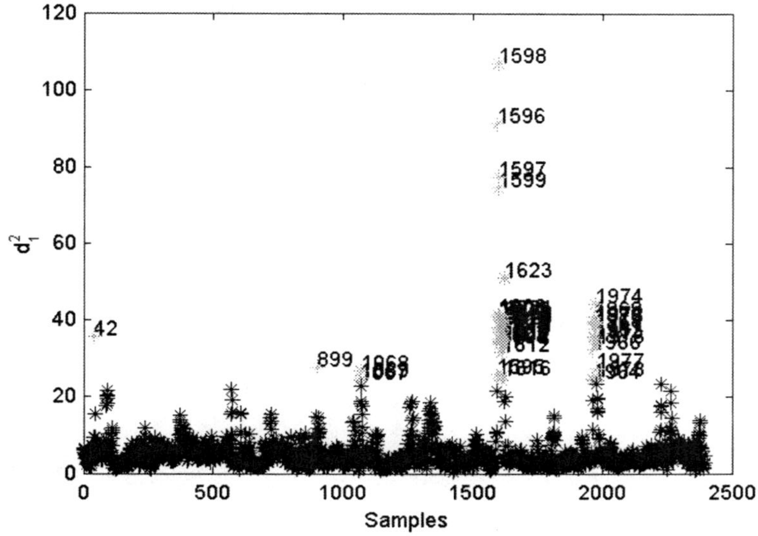

Figure 4.16. The first Jolliffe parameter (d_1^2) computed using PLS

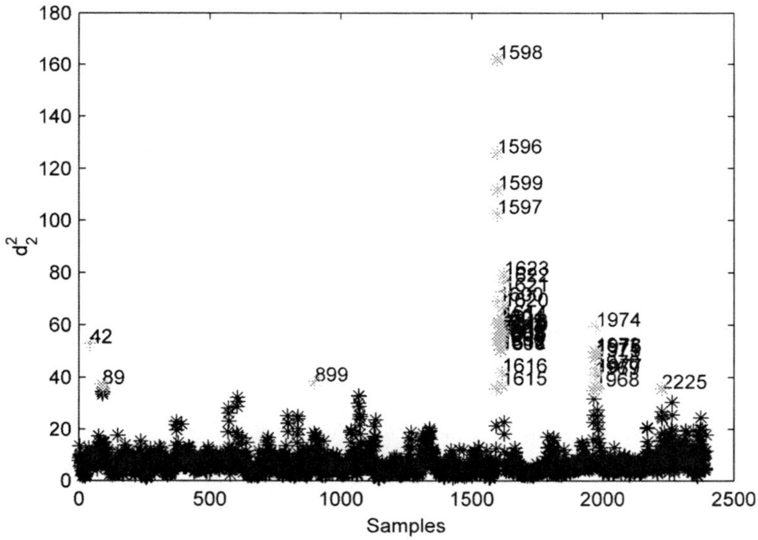

Figure 4.17. The second Jolliffe parameter (d_2^2) computed using PLS

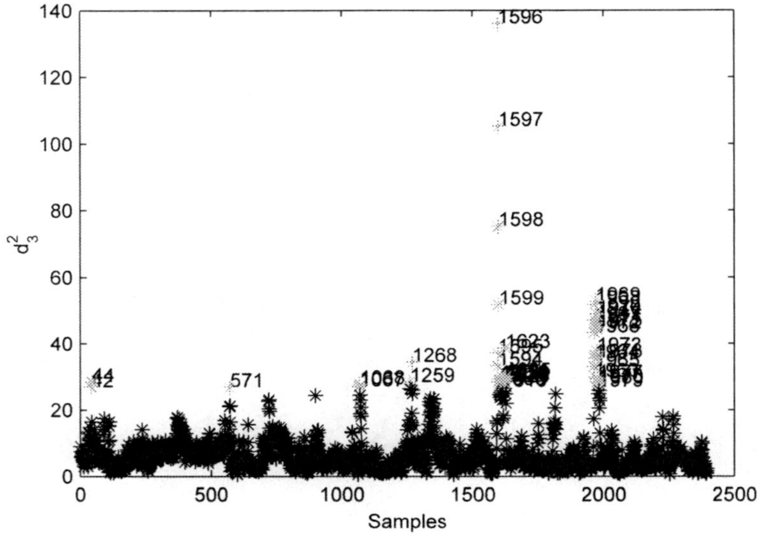

Figure 4.18. The third Jolliffe parameter (d_3^2) computed using PLS

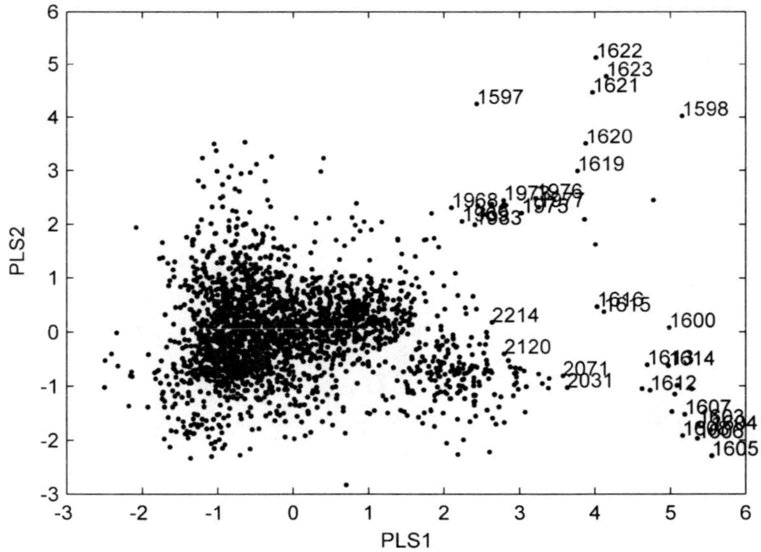

Figure 4.19. Plot of the first PLS component (PLS1) *vs.* the second PLS component (PLS2)

Also in this case, comparison with the plot of the first two PLS components can be useful. The results are shown in Figure 4.19, where labeled data once again refer to outliers suggested by all three Jolliffe parameters.

It is evident that in this case the region containing 'good' data is better defined than the corresponding region obtained using the PCA method. The drawback outlined above for PCA, *i.e.* the loss of data physical interpretation, holds also for this method.

4.1.4 Residual Analysis of Linear Regression

The final method described in this chapter consists in an analysis of the residual obtained after linear regression among independent and dependent variables. The same set of data taken from the debutanizer column database was considered in order to perform linear regression using the traditional LMS approach. A confidence level of 99.5% (corresponding approximately to 3σ in the hypothesis of Gaussian distribution of the residuals) was fixed as a threshold to recognize outliers in the residual vector. The residuals obtained are shown in Figure 4.20, along with the corresponding confidence interval. Outliers are reported in gray line color and correspond to data whose segments representing the confidence interval do not cross zero.

In Figure 4.21, a zoom of the data highlighted in the previous figure is reported, in order to better show the selection criterion for outliers.

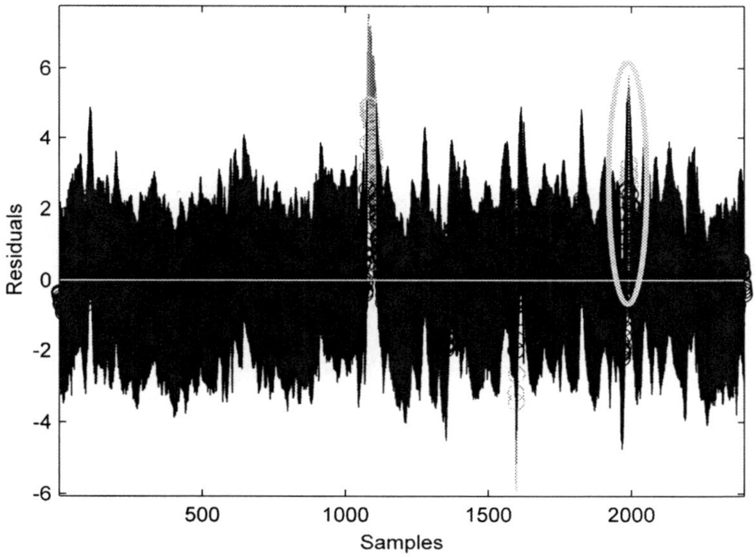

Figure 4.20. Time plot of the residual obtained after performing linear regression. The corresponding 3σ intervals are also reported. Outliers are reported in light gray color

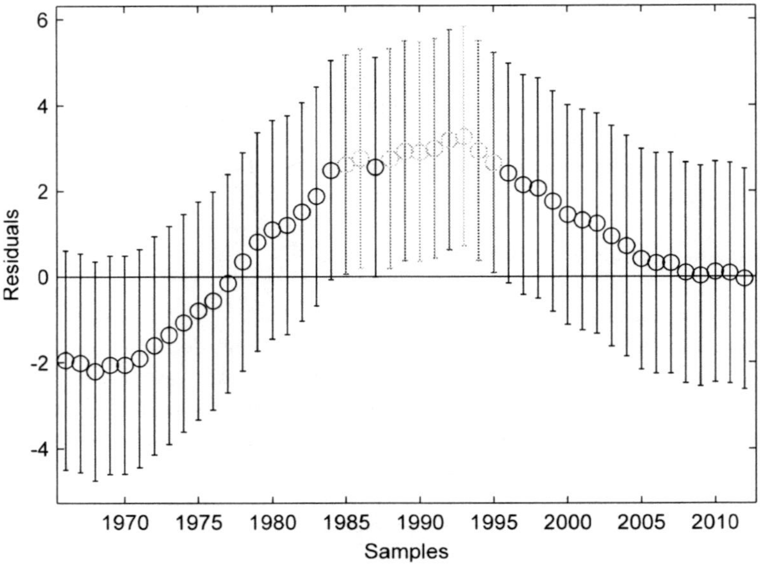

Figure 4.21. A zoom of the data reported in Figure 4.20: outliers correspond to segments that do not cross the zero line and are reported in gray color

4.2 Comparison of Methods for Outlier Detection

In this section, a comparative analysis of results obtained in the previous section is reported. The number of outliers detected by each method is summarized in Table 4.1.

Table 4.1. Outliers obtained for the debutanizer column with different detection strategies

		No. of outliers
3σ edit rule		233/2394
Jolliffe with PCA	d_1^2	40/2394
	d_2^2	44/2394
	d_3^2	42/2394
Jolliffe with PLS	d_1^2	50/2394
	d_2^2	48/2394
	d_3^2	44/2394
Linear regression analysis		45/2394

A first comparison is made between the number of outliers detected by the Jolliffe with PCA and Jolliffe with PLS methods. The results are reported in Table 4.2.

Table 4.2. Number of outliers in common between the Jolliffe with PCA and Jolliffe with PLS

		No. of outliers	No. of outliers in common	
Jolliffe with PCA	d_1^2	40	34	20
	d_2^2	44		
	d_3^2	42		
Jolliffe with PLS	d_1^2	50	31	
	d_2^2	48		
	d_3^2	44		

Comparison of the remaining methods, two at a time, are reported in Tables 4.3 to 4.7. In particular, the Jolliffe methods are compared by considering the outliers obtained by taking the three parameters together. In the tables, only the number of outliers in common are reported, while, at the end of the chapter, Table 4.8 reports the order number of the outliers recognized by each method. By using the data downloaded from the book web site, this table allows us to analyze, using time plots of the system variables, the characteristics of the outliers detected.

Table 4.3. Number of outliers in common between the Jolliffe with PCA and the 3σ edit rule

	No. of outliers	No. of outliers in common
Jolliffe with PCA	34	33
3σ edit rule	233	

Table 4.4. Number of outliers in common between the Jolliffe with PLS and the 3σ edit rule

	No. of outliers	No. of outliers in common
Jolliffe with PLS	31	31
3σ edit rule	233	

Table 4.5. Number of outliers in common between the 3σ edit rule and linear regression

	No. of outliers	No. of outliers in common
3σ edit rule	233	42
Linear regression	45	

Table 4.6. Number of outliers in common between the Jolliffe with PCA and linear regression

	No. of outliers	No. of outliers in common
Jolliffe with PCA	34	3
Linear regression	45	

Table 4.7. Number of outliers in common between the Jolliffe with PLS and linear regression

	No. of outliers	No. of outliers in common
Jolliffe with PLS	31	3
Linear regression	45	

From a critical analysis of the results reported in Tables 4.1 to 4.7 and by performing a visual inspection of time plots of the variables involved, reported in Subsection 4.1.1, it is possible to draw some conclusions about the process of outlier detection. In particular, most of the outliers suggested using the 3σ edit rule actually belong to relevant peaks of the output variable. Considering that such peaks are frequently repeated and that plant technologists required a model able to accurately estimate such dynamics, these values were not considered as outliers when modeling the debutanizer column.

The use of Jolliffe parameters, both in conjunction with PCA and PLS, generally gave consistent results. Differences between the two methods can be considered as a consequence of the inclusion of information about system output when the PLS is used.

The linear regression method gave results that generally do not fit with the suggestions afforded by other methods.

Table 4.8. List of outliers detected by each method

Samples	$\pm 3\sigma$	d_1^2 PCA	d_2^2 PCA	d_3^2 PCA	d_1^2 PLS	d_2^2 PLS	d_3^2 PLS	Linear regr.
42	•	•	•	•	•	•	•	
44							•	
88	•	•	•	•				
89	•	•	•	•		•		
90	•	•	•	•				
91	•	•	•	•				
92	•	•	•	•		•		
93	•	•	•			•		
369	•							
370	•							
371	•							
372	•							
570	•			•				
571	•	•	•	•			•	

Table 4.8. (continued)

Samples	±3σ	d_1^2 PCA	d_2^2 PCA	d_3^2 PCA	d_1^2 PLS	d_2^2 PLS	d_3^2 PLS	Linear regr.
572	•	•	•	•				
573	•	•	•	•				
574				•				
604	•			•				
605	•	•	•	•				
606		•	•	•				
721	•							
722	•							
796	•							
797	•							
798	•							
799	•							
833	•							
834	•							
835	•							
836	•							
837	•							
899	•	•	•	•	•	•		
934	•							
935	•							
1067	•				•		•	
1068	•				•		•	
1069	•				•			
1070	•							
1071	•							
1072	•							
1073	•							
1074	•							
1077	•							•
1078	•							•
1079	•							•
1080	•							•
1081	•							•
1082	•							•
1083	•							•
1084	•							•
1085	•							•
1086	•							•
1087	•							•
1088	•							•
1089	•							•

Table 4.8. (continued)

Samples	±3σ	d_1^2 PCA	d_2^2 PCA	d_3^2 PCA	d_1^2 PLS	d_2^2 PLS	d_3^2 PLS	Linear regr.
1090	•							•
1091	•							•
1092	•							•
1093	•							•
1094	•							•
1095	•							•
1096	•							•
1097	•							•
1098	•							•
1099	•							•
1100	•							•
1101	•							•
1102	•							•
1103	•							•
1104	•							•
1105	•							•
1106								•
1107								•
1108								•
1110	•							
1111	•							
1112	•							
1115	•							
1121	•							
1128	•							
1129	•							
1130	•							
1131	•							
1132	•							
1133	•							
1134	•							
1259							•	
1268							•	
1328	•							
1329	•							
1330	•							
1331	•							
1332	•							
1333	•							
1334	•							
1335	•							

Table 4.8. (continued)

Samples	±3σ	d_1^2 PCA	d_2^2 PCA	d_3^2 PCA	d_1^2 PLS	d_2^2 PLS	d_3^2 PLS	Linear regr.
1336	●							
1337	●							
1338	●							
1339	●							
1340	●							
1341	●							
1342	●							
1343	●							
1508	●							
1509	●							
1510	●							
1511	●							
1512	●							
1513	●							
1514	●							
1515	●							
1594	●						●	
1595	●		●		●	●	●	
1596	●	●	●	●	●	●	●	
1597	●	●	●	●	●	●	●	●
1598	●	●	●	●	●	●	●	●
1599	●	●	●	●	●	●	●	●
1600	●			●	●	●	●	
1601	●				●	●	●	
1602	●				●	●		
1603	●				●	●		
1604	●				●	●	●	
1605	●				●	●	●	
1606	●				●	●		
1607	●				●	●	●	
1608	●				●	●		
1609	●				●	●		
1610	●				●	●	●	
1611	●				●	●		
1612	●				●	●		
1613	●				●	●		
1614	●			●	●	●	●	
1615	●				●	●		
1616	●				●	●		
1617	●	●	●	●	●	●		
1618	●	●	●	●	●	●	●	
1619	●	●	●	●	●	●		

Table 4.8. (continued)

Samples	±3σ	d_1^2 PCA	d_2^2 PCA	d_3^2 PCA	d_1^2 PLS	d_2^2 PLS	d_3^2 PLS	Linear regr.
1620	•	•	•	•	•	•	•	
1621	•	•	•	•	•	•	•	
1622	•	•	•	•	•	•	•	
1623	•	•	•	•	•	•	•	
1624	•							
1625	•						•	
1816	•							
1963							•	
1964	•				•		•	
1965	•				•		•	
1966	•		•		•		•	
1967	•		•		•	•	•	
1968	•	•	•		•	•	•	
1969	•	•	•		•	•	•	
1970	•	•	•	•	•	•	•	
1971	•	•	•	•	•	•	•	
1972	•	•	•	•	•	•	•	
1973	•	•	•	•	•	•	•	
1974	•	•	•	•	•	•	•	
1975	•	•	•	•	•	•	•	
1976	•	•	•	•	•	•	•	
1977	•	•	•	•	•	•	•	
1978	•	•	•	•	•	•	•	
1979	•	•	•	•	•	•	•	
1980	•	•	•				•	
1981	•		•					
1982	•		•					
1983	•		•					
1984	•							
1985	•							•
1986	•							•
1987	•							
1988	•							•
1989	•							•
1990	•							•
1991	•							•
1992	•							•
1993	•							•
1994	•							•
1995	•							•
2030	•							
2088	•							

Table 4.8. (continued)

Samples	±3σ	d_1^2 PCA	d_2^2 PCA	d_3^2 PCA	d_1^2 PLS	d_2^2 PLS	d_3^2 PLS	Linear regr.
2139	•							
2162	•							
2222	•							
2223	•							
2224	•							
2225	•	•	•	•		•		
2226	•							
2227	•							
2235	•							
2262	•							
2263	•							
2264	•							
2265	•							
2266	•							
2267	•							
2277	•							
2278	•							
2279	•							
2280	•							
2281	•							
2282	•							
2283	•							
2284	•							
2285	•							
2286	•							
2287	•							
2288	•							
2289	•							
2290	•							
2291	•							
2292	•							
2293	•							
2294	•							
2295	•							
2296	•							
2297	•							
2298	•							
2299	•							
2300	•							
2301	•							
2302	•							
2303	•							

Table 4.8. (continued)

Samples	±3σ	d_1^2 PCA	d_2^2 PCA	d_3^2 PCA	d_1^2 PLS	d_2^2 PLS	d_3^2 PLS	Linear regr.
2304	•							
2305	•							
2306	•							
2307	•							
2308	•							
2309	•							
2310	•							
2311	•							
2312	•							
2313	•							
2314	•							
2315	•			•				
2316	•	•		•				
2317		•		•				
2344	•							
2345	•							
2346	•							
2347	•							
2348	•							
2349	•							
2350	•							

4.3 Conclusions

From authors experience, confirmed also from applications not reported here, it can be concluded that the task of outlier detection is a process that cannot be totally automated and all available information, including expert knowledge, has to be taken into account. Moreover, a fundamental role in the choice of outlier detection criterion is played by the final objective of the analysis.

As a general rule if a careful modeling of system dynamics is of interest, methods that exploit only statistical inspection of process data can have an undesired conservative effect. They tend to remove peaks which can carry precious information about system dynamics. A better choice in these cases is methods like PLS that take into account input–output relationships. The results obtained should be in any case carefully investigated due to the rough nature of the outlier detection process. The suggested methods, in fact, limit the search process to the case of linear static correlations, while their nonlinear extensions are usually computationally expensive. If a coarse model is of interest, a rule of thumb is that in the case of an uncertain candidate, it is better to eliminate it, instead of using wrong information in the process of model identification. Of course the number of available data can make the designer more or less sensitive to this rule.

5

Choice of the Model Structure

5.1 Introduction

This chapter describes a number of case studies referring to soft sensor design for systems that require different model structures. We begin with static linear and nonlinear models, then move on to dynamic models. Accordingly, a number of strategies will be used for the selection of model inputs, spanning from a trial and error approach to analytical approaches based on correlation analysis, Mallows coefficients, and PLS.

Though a number of different approaches will be considered that can help model selection, it should be borne in mind that the available methods are generally designed for linear models or else are an extension of strategies, developed for linear models, to the case of nonlinear models. In fact, most methods are based on the analysis of some kind of correlation between variables, and correlation is indeed an indicator of linear dependence.

In Section 5.2 the case of a static model for pollutant concentrations in fumes is approached using both linear and nonlinear models.

The estimation of research octane number (RON) for powerformed gasoline (*i.e.* gasoline with a higher octane number) is used to discuss linear dynamic models in Section 5.3 (this same case study will be reconsidered in Chapter 7, where the fuzzy aggregation of two nonlinear models is designed). Different nonlinear modeling strategies are compared in Section 5.4, using a sulfur recovery unit (SRU) as case study. Finally, in Section 5.5 a number of approaches designed to help with the influencing variables and regressor choices for NMA models are described, with reference to a debutanizer column.

5.2 Static Models for the Prediction of NO_x Emissions for a Refinery

In this section, static models designed to estimate the NO_x emissions produced by chimneys in a large refinery are proposed. Details about the plant are reported in the Appendix. For this process, a model is required to replace the on-line analyzer during maintenance. In fact, the plant is equipped with a gas chromatograph that measures the NO_x level with a sampling time T_s = 1 min. Due to the harsh environment, the analyzer is frequently off-line for maintenance.

Data produced by the analyzer are collected in the refinery database and subsequently used to compute the monthly average value. In particular, Italian laws establish a limit on the monthly average emission level of NO_x, com puted on the base of hourly recorded data. Moreover, the average value is considered valid if the minimum percentage of acquired hourly data is 80% or higher. In the event that such a percentage is not available, *e.g.* due to failure of on-line analyzers, Italian laws require the use of mathematical models that estimate the NO_x level on the basis of chimney inputs. A linear model, developed on the basis of adequate sensitivity coefficients, and suggested by the refinery experts' heuristic knowledge, is used in this section as a benchmark, while a comparison of different linear and nonlinear static data-driven models is reported later on in the section. In particular, performances obtained by using a neural nonlinear model are compared with the estimation capability of both the heuristic model and a linear model, obtained using a LMS approach.

As reported in the Appendix, input variables are represented by the flows of a number of different fuel oils and gases, feeding processes whose fumes are conveyed in the chimney.

The data considered were obtained using records produced by the gas chromatograph during a period lasting about three months. A total number of some 1900 valid samples was considered, each one representing a mean hourly value, in accordance with Italian regulations.

The empirical model is

$$y = \sum_{i=1}^{9} w_i \, u_i \tag{5.1}$$

where y is the concentration of NO_x in the chimney fumes and u_i the ith group of the chimney input. In particular, the total number of process inputs is 18, grouped into 9 inputs, in accordance with a relative importance weight w_i determined by process experts on the basis of heuristic knowledge. In Figures 5.1, 5.2 and 5.3 the performance of the heuristic model is shown. In Figure 5.1, the model output is compared with data obtained from gas chromatograph acquisitions. In Figures 5.2 and 5.3, the corresponding residual and the residual histogram are reported.

It can be observed that the heuristic model generally overestimates the NO_x emission level. Though this characteristic is conservative as regards environmental pollution, more efficient estimations are of interest to the refinery operators.

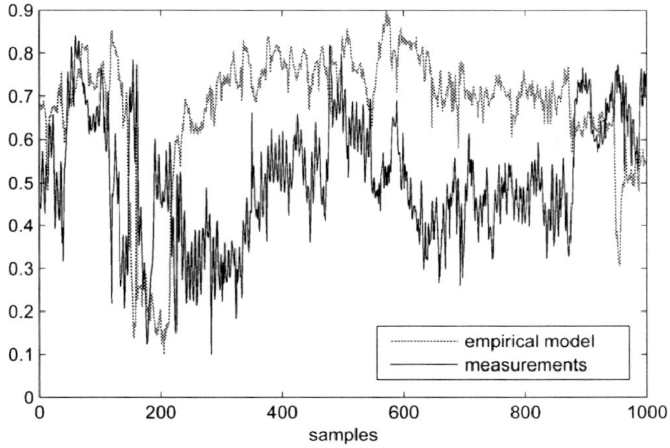

Figure 5.1. Comparison between measurements and the empirical model output, normalized validation data

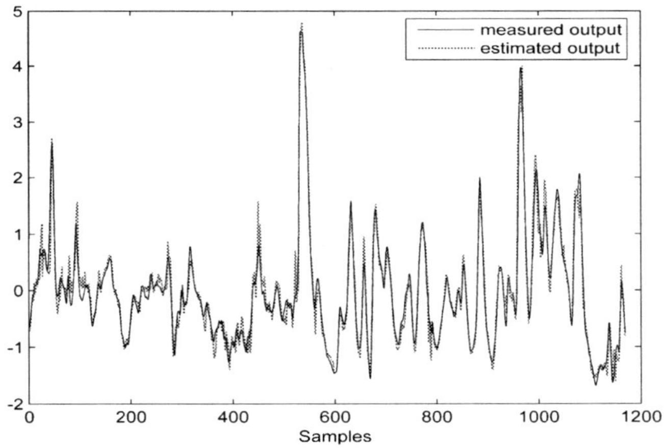

Figure 5.2. Empirical model residual

An improvement of the prediction performance of the linear model given in Equation 5.1 was obtained when computing the model coefficients using the LMS approach. In this case the original 18 inputs were taken into account separately and the available data set was split in order to obtain both a training and a validation data set. The results reported below refer to the performance of the model evaluated on the validation data set.

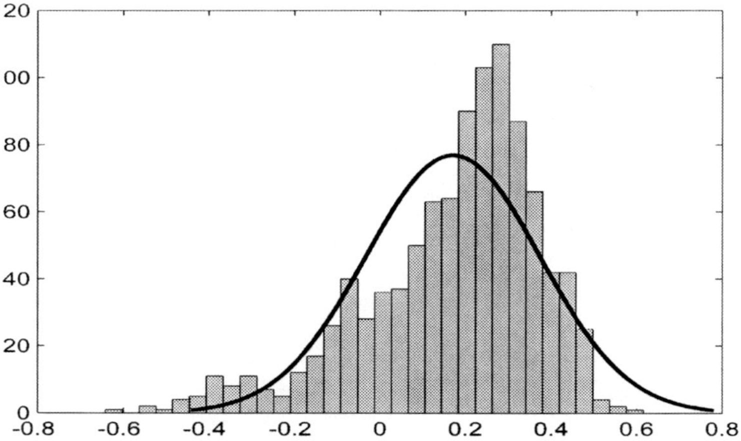

Figure 5.3. Empirical model residual histogram

In Figure 5.4, the performance of the LMS linear model with 18 inputs is compared with measured data, and in Figures 5.5 and 5.6 the corresponding residual and residual histogram are reported.

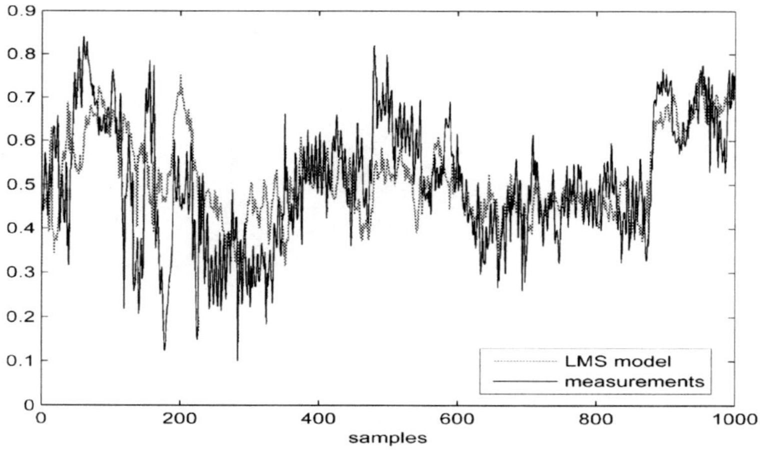

Figure 5.4. Comparison between measurements and the LMS-based model output, normalized validation data

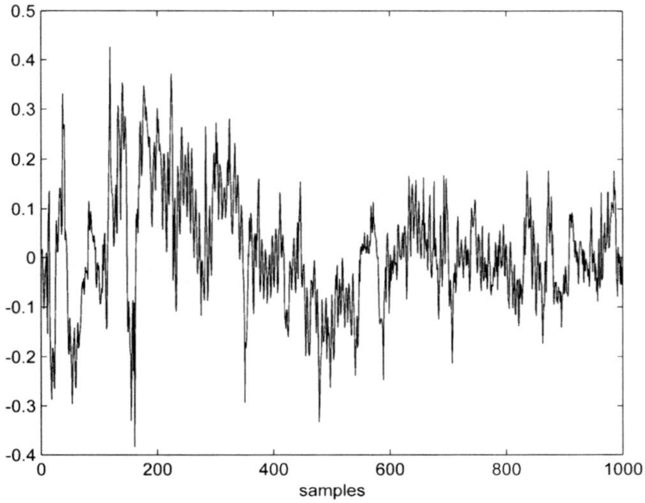

Figure 5.5. LMS model residual

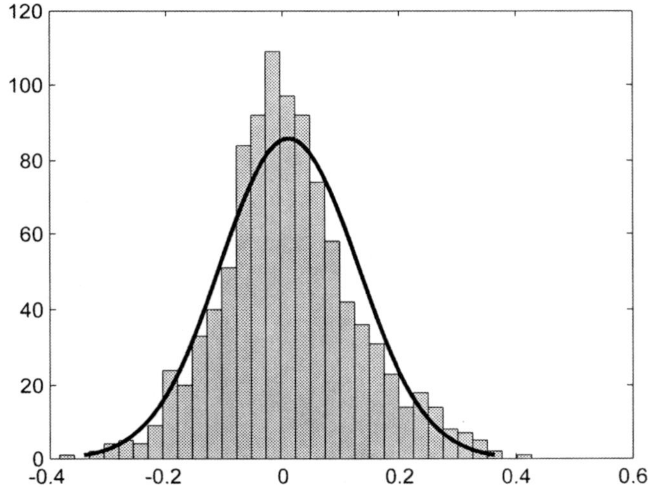

Figure 5.6. LMS model residual histogram

It can be observed that the linear LMS model outperforms the heuristic model even if the model performance can be further improved using nonlinear models. Neural models, based on MLP networks, were designed using the same input variables and data sets. One hidden layer 18-11-1 MLP, trained using the Levenberg–Marquardt algorithm with the early stopping strategy, gave the best results.

Figures 5.7, 5.8 and 5.9 show the results obtained.

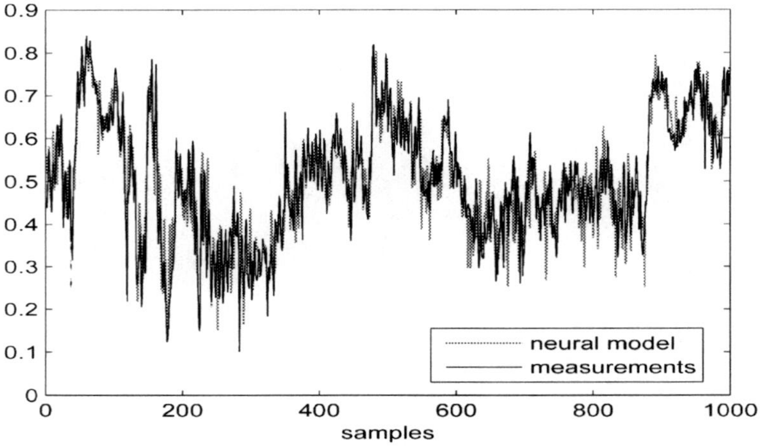

Figure 5.7. Comparison between measurements and the neural model output, normalized validation data

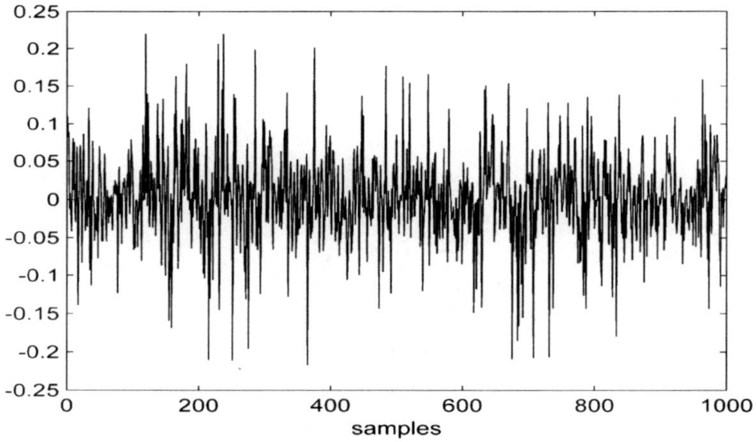

Figure 5.8. Neural model residual time plot

In Table 5.1 the correlation coefficients between measured data and the corresponding estimations, obtained with the models considered, for the validation data set are reported. It can be noted that the linear LMS model works better than the heuristic model and that the nonlinear MLP based model offers the best performance in terms of correlation between real data and estimated values.

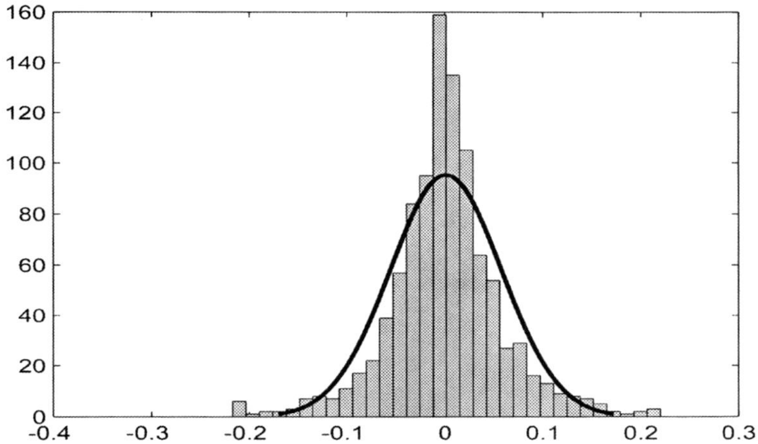

Figure 5.9. Neural model residual histogram

Table 5.1. Correlation coefficients between real data and estimated values and standard deviation of model residals

	Empirical model	LMS model	Neural model
Correlation coefficient	0.02	0.66	0.93
Residual standard deviation	0.21	0.11	0.06

5.3 Linear Dynamic Models for RON Value Estimation in Powerformed Gasoline

Linear models should be tried first when a soft sensor is searched for. In fact, this class of models has been investigated in depth and many theoretical results on both design and validation phases are available. Moreover, linear models are more suitable for developing a control system, and even if more complex structures will be the final design result, they represent a useful benchmark. These interesting aspects are greatly overshadowed by the nonlinear nature of most industrial processes. Hence the approximation capability of linear models is frequently unsatisfactory.

In this section, an example of a soft sensor based on a linear dynamic model is introduced. The possibility of estimating the RON in gasoline produced by a powerformer unit, using this class of models, is investigated. The estimation was required by the plant technologists in order to replace on-line measuring devices during planned maintenance actions. MA models are therefore of interest also in this case.

In the application considered, two different working conditions can easily be recognized, depending on one of the input variables. In Figure 5.10 a time plot of this variable showing large jumps between two ranges of values is reported. These jumps force the plant to move from one working point to the other.

For this reason, as a first strategy, two different linear dynamic models were designed; the performance of these models is the subject of this section, while in a subsequent chapter a strategy to improve soft sensor performance will be described.

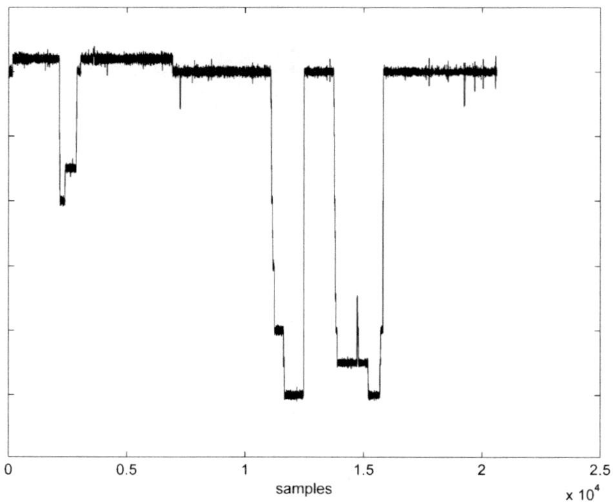

Figure 5.10. The input variable of the powerformer unit, used to discriminate the two working points of the plant

Data used to derive the models were collected during a period lasting about six weeks with a sampling period of 3 min (corresponding to about 20000 samples, after invalid data elimination). More details about the plant and the variables involved can be found in the Appendix.

As usual, the first step in model identification consists in the choice of input variables. The experts suggested that six variables should be used as model inputs.

Also, a suitable number of regressors were fixed for each input quantity, on the basis of both correlation analysis and considerations regarding system physics; the following model structure resulted:

$$\begin{aligned} y(k) &= H + a_1u_1(k) + a_2u_1(k-3) + a_3u_1(k-5) + b_1u_2(k) \\ &+ b_2u_2(k-5) + c_1u_3(k) + c_2u_3(k-4) + c_3u_3(k-5) + d_1u_4(k) \\ &+ d_2u_4(k-5) + e_1u_5(k) + e_2u_5(k-1) + f_1u_6(k) + f_2u_6(k-1) \end{aligned} \tag{5.2}$$

where $y(k)$ is the kth sample of the RON and $\{H, a_1, a_2, a_3, b_1, b_2, c_1, c_2, c_3, d_1, d_2, e_1, e_2, f_1, f_2\}$ is a set of multiplicative coefficients to be searched for, that was estimated using the classical LMS approach.

Simulation results on a set of validation data are shown in Figure 5.11, while in Figure 5.12 the 4-plot analysis of the residual is reported.

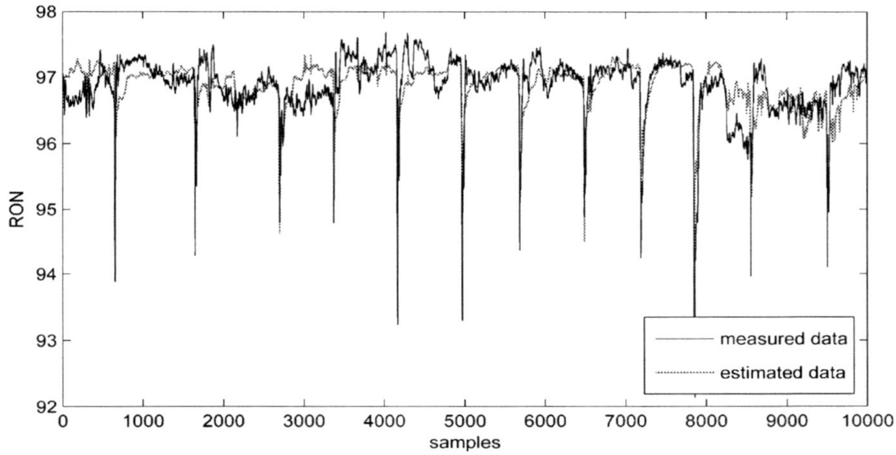

Figure 5.11. Acquired RON values and their linear estimation, scaled units

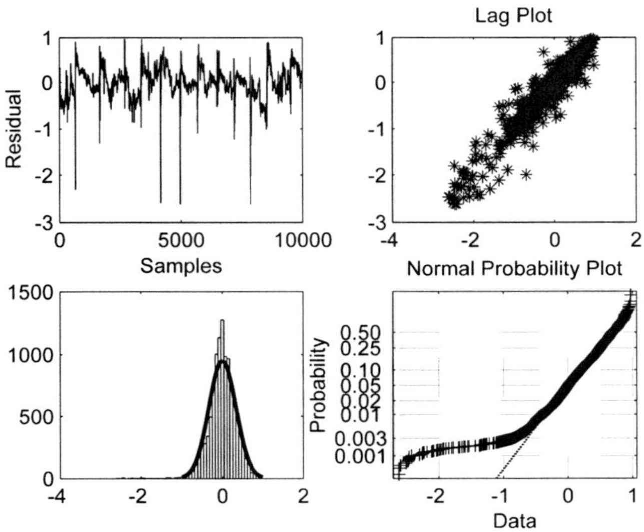

Figure 5.12. 4-plot of the residual of the linear model

Though the linear model follows the real-time series trend, the model accuracy was considered not satisfactory by the plant technologists, who considered this to be a consequence of the plant nonlinearity, which cannot be modeled using a linear model structure.

5.4 Soft Computing Identification Strategies for a Sulfur Recovery Unit

In this section, a study case is reported dealing with comparison of the performance obtained using different nonlinear structures to realize a soft sensor. NMA models are implemented and compared with respect to both performance and ease of implementation. The strategies used were MLP and radial basis function (RBF) neural networks, neuro-fuzzy (NF) networks and nonlinear least square fitting (NLSQ).

As a design guideline, only NMA models were considered. This kind of model is particularly suitable for a soft sensor to replace an actual sensor for a long period, because it does not require past samples of the output. Alternatively, autoregressive models fed with past predicted output values can be adopted. In that case, a very accurate estimation capability is needed to minimize the effect of estimate error propagation. Nonlinear autoregressive models were designed and tested for the application described in this work; however, their performance was poorer than that obtained with NMA models.

Regressors of the input variables are dealt with in this section using a trial and error procedure. However, the possibility of using automatic procedures for regressor selection is described in great detail in other sections of this chapter.

The soft sensors described in this section measure the concentration of the so called acid gases, hydrogen sulfide (H_2S) and sulfur dioxide (SO_2), in the tail stream of a Sulfur Recovery Unit (SRU).

Soft sensors are designed to compute the SRU tail gas composition on-line, by exploiting a suitable set of measurements of the input variables of the process. In particular, the soft sensors presented consist of nonlinear dynamic models, capable of predicting the concentration of H_2S and SO_2 separately. In what follows, a complete analysis of results for H_2S are reported. Similar results were obtained for SO_2 estimation.

In this application the learning set includes 1000 samples, with a sampling time of 1.0 min. A second set of 1000 samples was selected as a validation set, to prevent overfitting via early stopping of the learning phase when designing the neural models. A further, larger, set was used as a checking data set. Plant and input variables description are reported in the Appendix.

As a starting point, the work by Quek, Balasubramanian and Rangaiah (2000) was considered, and a static nonlinear model was realized using a MLP. The modeling results were not satisfactory, especially as regards the peak value estimates, which the experts consider to be the most important in this application.

Subsequently, NMA models were taken into account. Some choices had to be made about the regressors. As stated above, delayed samples were selected during a trial and error phase. The experience of plant operators, the correlation coefficients between delayed replicas of the inputs and each output, and the performance of the model itself were considered. The chosen structure was therefore used to design a number of nonlinear models using the same data sets. This strategy was adopted because the main objective of this section is to compare different nonlinear implementations of soft sensors.

The model structure adopted is the following:

$$y(k) = \begin{matrix} f(x_1(k), x_1(k-5), x_1(k-7), x_1(k-9),\ldots \\ \ldots, x_5(k), x_5(k-5), x_5(k-7), x_5(k-9)) \end{matrix} \tag{5.3}$$

The results obtained are shown below in graphical form. In particular, for each modeling strategy a comparison between scaled measured data and their estimations is reported both for the training data set and for a relevant subset of the testing data set.

Figures 5.13 and 5.14 show the results obtained using a 20-8-1 MLP, trained with the Levenberg–Marquardt algorithm with early stopping strategy.

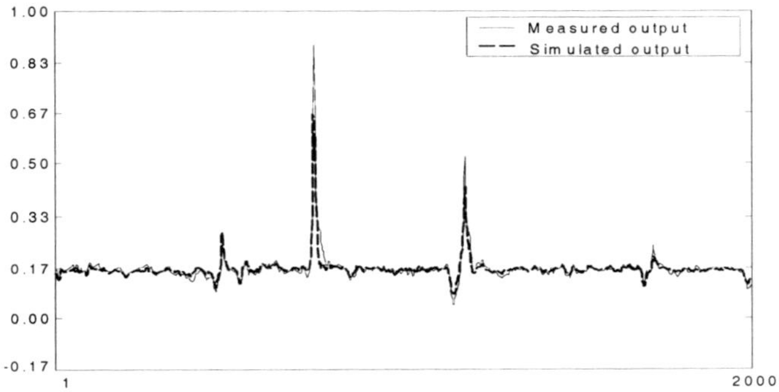

Figure 5.13. Comparison between simulated and measured H_2S values for the neural model, scaled training data

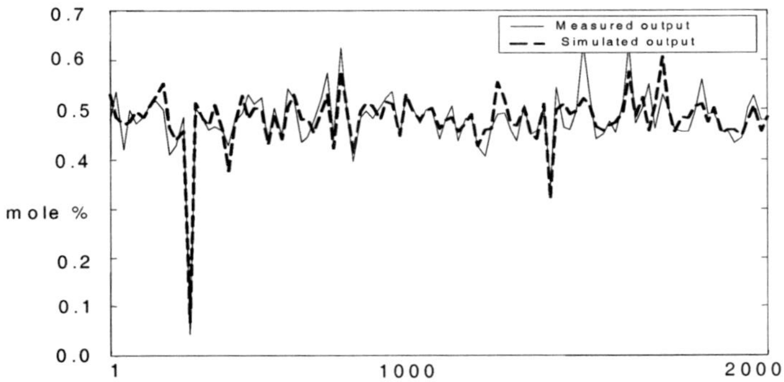

Figure 5.14. Comparison between simulated and measured H_2S values for the neural model, subset of scaled validation data

A second approach to implement Equation 5.3 was based on RBF networks (Chen, Cowan and Grant, 1991). RBF networks were created iteratively by adding one neuron at a time until a satisfactory value of the estimation mean square error was reached. The best results with RBF networks were obtained with 298 neurons. Although the RBF-based model gave the best performance on the learning patterns, its performance on the checking patterns was very poor. Better results were obtained only with a great increase in the number of neurons, but this strategy was not pursued because of the corresponding complexity of an eventual on-line implementation.

The third strategy considered was based on NF models. In particular, the ANFIS (adaptive neuro-fuzzy inference system), implemented in the Fuzzy Logic Toolbox (Matlab®) was used to implement the NMA model. ANFIS is based on the Takagi–Sugeno–Kang inference method (Takagi and Sugeno, 1985), in which the output membership functions in the consequent of each rule are singleton spikes. Parameters associated with membership functions are chosen to tailor the membership function to the I/O data via a neuro-adaptive learning technique based on Fuzzy C-mean clustering (Bezdek, 1981) and a backpropagation algorithm. Based on Equation 5.3, the i_{th} rule of the FIS has the following form:

$$\textit{if } \text{in}_1 \textit{ is } \text{MF}_1 \textit{ and } \text{in}_2 \textit{ is } \text{MF}_2 \textit{ and} \dots \text{in}_{20} \textit{ is } \text{MF}_{20} \textit{ then } \text{y} \textit{ is } \text{z}_i \tag{5.4}$$

The best configuration of the FIS model consists of six rules and six membership functions per input. An attempt to reduce the number of membership functions was performed manually, by fusion of very similar membership sets, leading to four fuzzy membership sets per input and six rules, without affecting modeling performance. As an example, in Figure 5.15 the membership functions obtained for variable in_5 are given.

The results are illustrated in Figures 5.16 and 5.17.

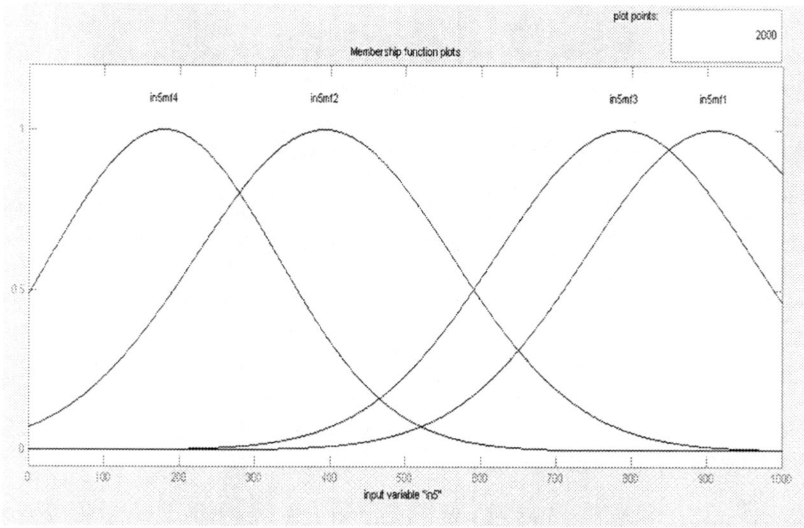

Figure 5.15. Comparison between the membership functions for variable in_5 obtained after manually pruning the FIS model

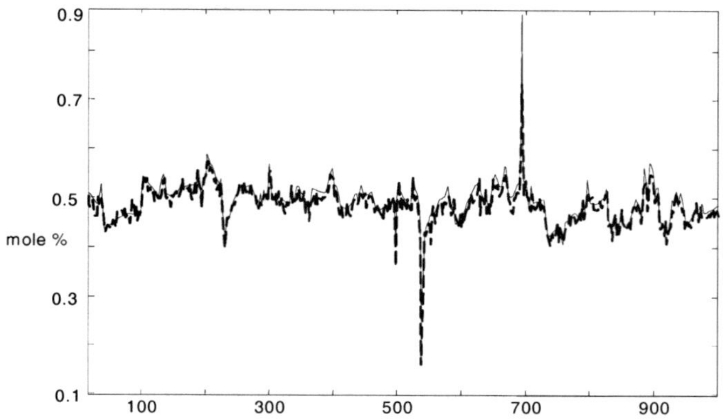

Figure 5.16. Simulated and measured H_2S comparison of the model implemented via a neuro-fuzzy network, scaled learning data

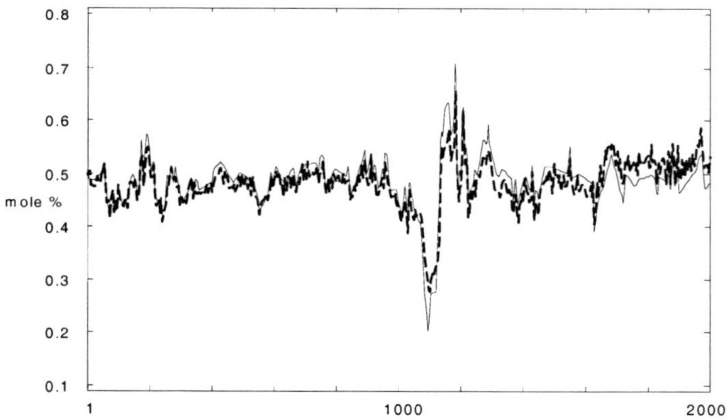

Figure 5.17. Simulated and measured output H_2S comparison of the model implemented via a neuro-fuzzy network, subset of scaled validation data

The H_2S model was also implemented by exploiting a nonlinear multivariable rational function, like that used in Quek, Balasubramanian and Rangaiah (2000) on a static model of a similar SRU plant. In that work, a second-order nonlinear function was adopted to implement the dynamic NMA model. With this approach function $f(\cdot)$ in Equation 5.3 has the following form:

$$y(k) = \frac{a_0 + a_1 x_1(k) + a_2 x_1^2(k) + \cdots + a_{19} x_5(k-9) + a_{20} x_5^2(k-9)}{b_0 + b_1 x_1(k) + b_2 x_1^2(k) + \cdots + b_{19} x_5(k-9) + b_{20} x_5^2(k-9)} \tag{5.5}$$

The coefficients were computed using a nonlinear least square data fitting algorithm (Ponton and Klemes, 1993). The results are illustrated in Figures 5.18 and 5.19.

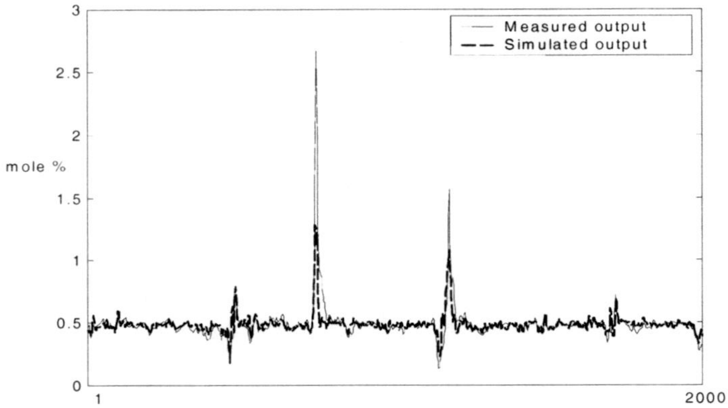

Figure 5.18. Comparison between simulated and measured H_2S for the nonlinear LSQ-based model, learning and validation data sets

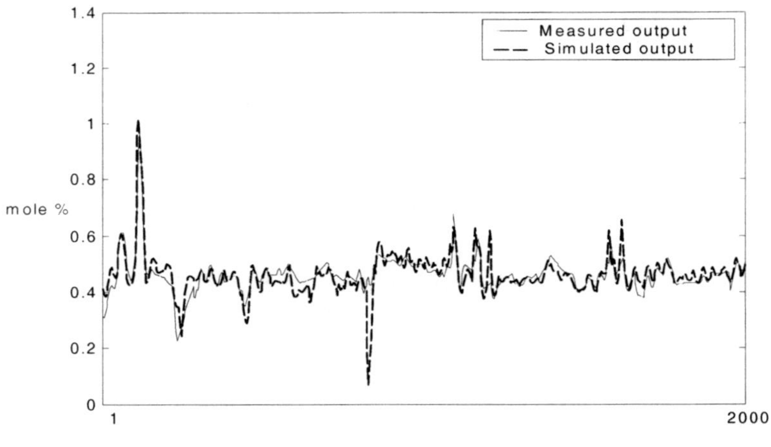

Figure 5.19. Comparison between simulated and measured H_2S for the model implemented via nonlinear LSQ, subset of scaled validation data

As mentioned above, a similar procedure was adopted to model the concentration of SO_2.

Some comparisons were made by visual inspection of graphs like the reported ones, MSE and correlation coefficients. Tables 5.2 and 5.3 give, respectively, the

MSE, computed for each model, on both the training and the checking data set, and the corresponding correlation coefficient between the simulated and computed output.

Table 5.2. MSE obtained with different implementations of the models

MSE	MLP	RBF	ANFIS	NLSQ
H_2S training Data	0.0008	0.0002	0.0009	0.0007
SO_2 training Data	0.0003	0.0002	0.0006	0.0004
H_2S checking Data	0.0009	0.0018	0.0012	0.0008
SO_2 checking Data	0.0004	0.0015	0.0008	0.0004

Table 5.3. Correlation coefficients between acquired data and their estimations, obtained with different implementations of the models

Correlation coefficient	MLP	RBF	ANFIS	NLSQ
H_2S training Data	0.851	0.939	0.843	0.858
SO_2 training Data	0.919	0.941	0.852	0.897
H_2S checking Data	0.847	0.722	0.813	0.848
SO_2 checking Data	0.903	0.761	0.865	0.905

A further element that should be taken into account when on-line implementation of a soft sensor is needed is its computational complexity. Table 5.4 compares the number of mathematical operations required by each of the models.

Table 5.4. Comparison of the number of operations for different implementation of models

	Sums	Products	Exponentials
MLP	176	176	8
ANFIS	130	365	120
NLSQ	80	121	0

From an analysis of the previous figures and tables, it can be noticed that RBF-based models offer the best performance with learning data, but are not able to confirm this performance on the checking set, where, instead, they give the worst results. They were therefore not considered anymore.

NF, NLSQ and MLP models give comparable satisfactory performance, with a slight improvement for NLSQ and MLP. Moreover, the NLSQ model requires the smallest number of operations.

Based on performance and computational complexity issues, both NLSQ and MLP were tested to verify their on-line performance: MLP-based models offered the best performance during the whole period of observation, whereas NLSQ revealed some periods of mismatching.

Examples of acquired data and of the corresponding on-line estimations, obtained with the MLP soft sensor, are reported in Figures 5.20 and 5.21 for H_2S and SO_2, respectively. Due to commercial confidentiality, graphs are reported without the vertical axis scale.

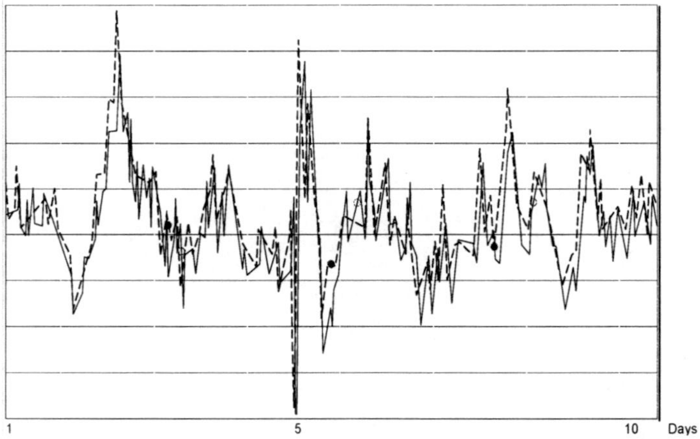

Figure 5.20. On-line performance of the model for the estimation of H_2S. Solid line refers to real data and dotted line to model estimation

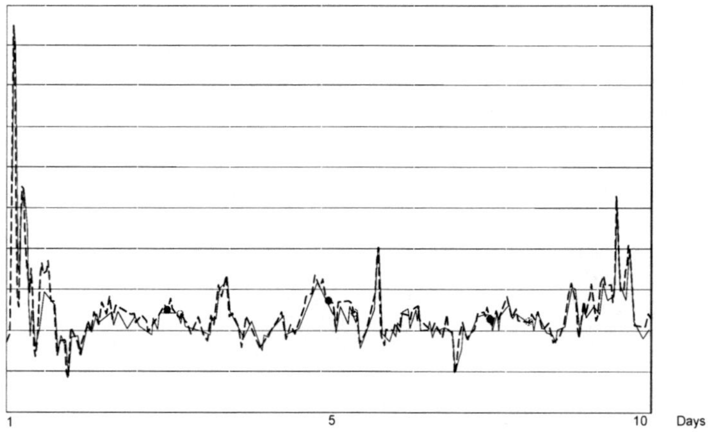

Figure 5.21. On-line performance of the model for the estimation of SO_2. Solid line refers to real data and dotted line to model estimation

From the case studies reported so far, it appears evident that the choice of design strategy plays a main role in the resulting soft sensor capability. Hence, though each designer tends to use just a few approaches, investigating an affordable number of strategies can result in significant performance improvements. Last but not least important, computational effort should be taken into account when the soft sensor is designed to work on-line and must be implemented using existing resources. In this case, the final choice is usually a compromise.

5.5 Comparing Different Methods for Inputs and Regressor Selection for a Debutanizer Column

In this section, the problems of relevant inputs and regressor selection for a dynamic system are addressed. A number of methods are applied using as case study the debutanizer column described in the Appendix.

Two quantities are of relevance for the plant experts: real-time estimation of both the C5 (stabilized gasoline) content in the overheads of the debutanizer column and the C4 (butane) content in the bottom flow to stock. Soft sensors can be used for this purpose (Fortuna, Graziani and Xibilia, 2005a).

The case of estimating the butane content in the bottom flow will be considered.

As reported in the Appendix, in the case considered the measuring system for the output of the plant is installed on a different column, and this introduces a long delay that does not allow the newest regressors of the measured output to be used (only values $y(k-\Delta)$ and older are available at discrete time k, where Δ belongs to the interval [20–60] min). For this reason, NMA models will be considered here, while solutions using the available output regressors together with a number of estimated output samples will be introduced later in the book.

The plant experts suggested a set of eight candidate input variables for the model and one dependent variable. Two different problems need to be solved: the selection of a minimum set of relevant input variables and the detection of the corresponding regressors. These two problems will be approached together by considering a starting set of candidate input variables built using three regressors for each plant variable. The applied methods, designed to select influential variables, in this case process all the 24 elements in the same way and therefore determine both model input variables and corresponding regressors.

Though the number of regressors considered could be thought too restrictive, this number is a compromise between the computational load of the proposed strategy and the relevant dynamics of the system. As usual, knowledge of the plant physics can help the user to accomplish this task avoiding wrong choices.

The methods taken into account are

- simple correlation method;
- partial correlation method;
- Mallow's coefficients with a linear model;
- Mallow's coefficients with a neural model;
- PLS-based methods.

Of course, these methods are not exhaustive and a number of other methods can be found in the literature, as reported in Chapter 1.

5.5.1 Simple Correlation Method

Graphical analysis of scatter plots of the plant output against each candidate input can give useful information about dominant linear and/or nonlinear dependencies.

In fact, in the case of a strong direct linear dependence it should be easy to recognize in the scatter plot, whereas a parabolic trend suggests a quadratic dependence and so forth. In Figure 5.22 the scatter plot of the debutanizer output variable at time *k* versus input variables at the same time is shown for a relevant set of recorded data, while scatter plots of the output versus input variables at time $k-4$ are shown in Figure 5.23. Similar graphs were obtained when the input variables at sampling time $k-8$ were taken into account.

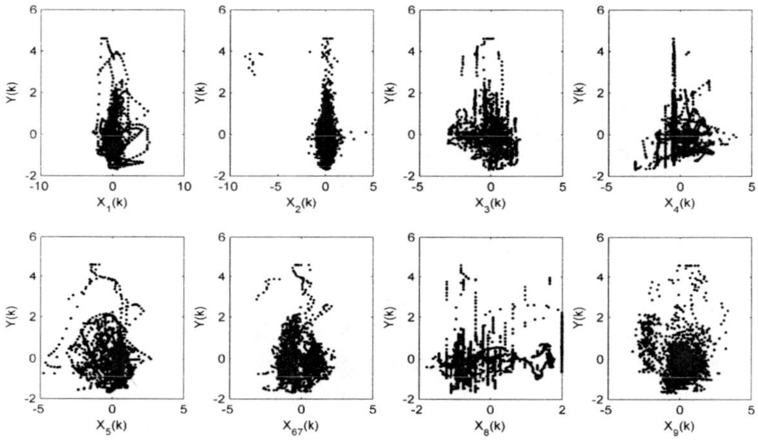

Figure 5.22. Scatter plot of the debutanizer output variable at time *k* versus input variables at the same time

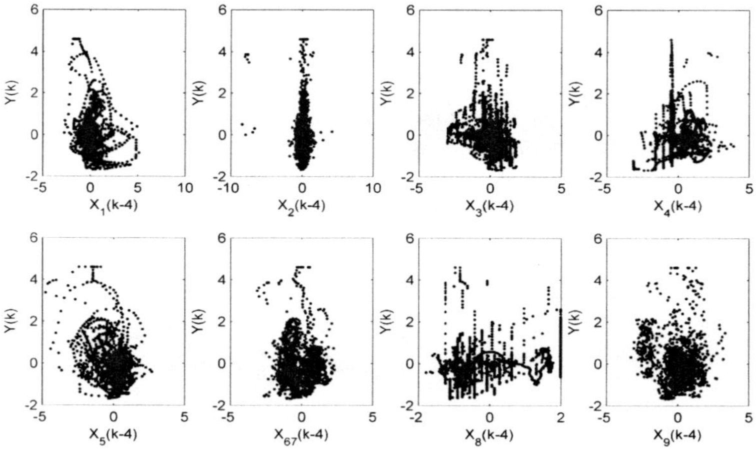

Figure 5.23. Scatter plot of the debutanizer output variable at time *k* versus input variables at time $k-4$

Although visual inspection of the scatter plots does not allow any direct input/output dependence to be established, it cannot be concluded that dependence does not exist. In fact, scatter plots cannot highlight any dependence among the system output on a number of input variables.

A quantitative criterion generally used to search for input–output relationships is the estimation of correlation coefficients between the system output and each candidate input, along with corresponding regressors. Though in this case a numerical indication of the correlation strength is obtained, only linear dependence can be recognized.

Table 5.5 illustrates the correlation coefficients between the debutanizer output and the selected set of inputs. For more details on the physical meaning of each variable in the model, the reader should refer to the Appendix. In particular, the variable named x_{67} is the mean value of input variables x_6 and x_7.

It can be observed that the absence of a clear linear dependence is reflected, for this indicator, in low values of the correlation coefficients. Nevertheless, if one is interested in testing the performance of a linear model, computation of the correlation coefficients can give useful information for the selection of the most relevant inputs. In fact, correlation coefficients give a quantitative selection criterion: if a linear model is of interest, the larger the correlation coefficient, the stronger are the correlation results, and this information can serve as a guide in the selection of model inputs.

In the case considered, Table 5.5 was analyzed in order to search for entries that gave the largest correlation coefficients. A set of nine inputs for the model was selected. The corresponding entries are represented in Table 5.5 by gray cells.

In Figure 5.24, correlation coefficients are plotted in order to have visual evidence of the selection criterion adopted, corresponding to a threshold value for the correlation coefficient of 0.2.

Table 5.5. Correlation coefficients between the output variable at time k versus input variables at times k, $k-4$ and $k-8$

	$x_1(k)$	$x_2(k)$	$x_3(k)$	$x_4(k)$	$x_5(k)$	$x_{67}(k)$	$x_8(k)$	$x_9(k)$
c_{xy}	0.0163	0.1684	0.2554	0.1416	0.1708	0.0100	0.2915	0.1041
	$x_1(k-4)$	$x_2(k-4)$	$x_3(k-4)$	$x_4(k-4)$	$x_5(k-4)$	$x_{67}(k-4)$	$x_8(k-4)$	$x_9(k-4)$
c_{xy}	0.1101	0.0950	0.2873	0.1393	0.3562	0.0538	0.2821	0.1325
	$x_1(k-8)$	$x_2(k-8)$	$x_3(k-8)$	$x_4(k-8)$	$x_5(k-8)$	$x_{67}(k-8)$	$x_8(k-8)$	$x_9(k-8)$
c_{xy}	0.2411	0.0108	0.2963	0.1080	0.5289	0.1189	0.2624	0.1508

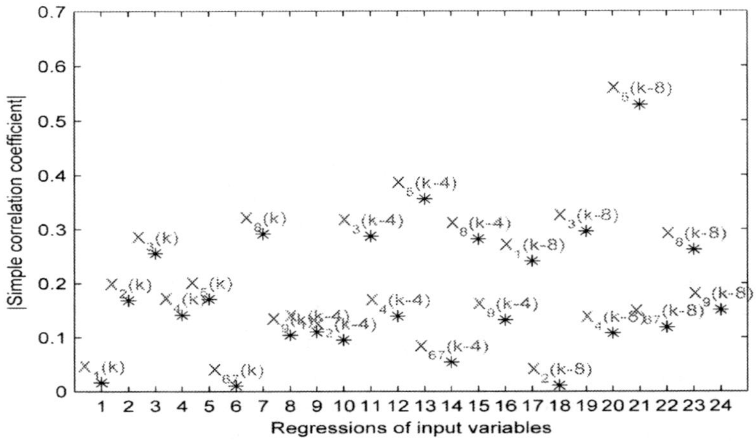

Figure 5.24. Simple correlation coefficients of the debutanizer output variable at time k versus input variables at times k, $k-4$ and $k-8$

5.5.2 Partial Correlation Method

A simple regression method can fail for linear models in those cases when the linear correlation between the output variable and an input one is overshadowed by the effects of other variables. In such an event, the partial correlation method can be helpful, as explained in Chapter 3.

Here this method is applied to the debutanizer in order compare the suggested model structure. In Table 5.6, the results obtained by applying the partial correlation method between the debutanizer column output and both the corresponding inputs and their regressors are reported.

Table 5.6. Partial correlation coefficients between the output variable at time k versus input variables at times k, $k-4$ and $k-8$

	$x_1(k)$	$x_2(k)$	$x_3(k)$	$x_4(k)$	$x_5(k)$	$x_{67}(k)$	$x_8(k)$	$x_9(k)$
c_{xy}	0.0935	0.1660	0.0356	0.0415	0.0728	0.1167	0.0115	0.0899
	$x_1(k-4)$	$x_2(k-4)$	$x_3(k-4)$	$x_4(k-4)$	$x_5(k-4)$	$x_{67}(k-4)$	$x_8(k-4)$	$x_9(k-4)$
c_{xy}	0.0169	0.0293	0.0317	0.0165	0.0526	0.0624	0.0496	0.0371
	$x_1(k-8)$	$x_2(k-8)$	$x_3(k-8)$	$x_4(k-8)$	$x_5(k-8)$	$x_{67}(k-8)$	$x_8(k-8)$	$x_9(k-8)$
c_{xy}	0.0026	0.0720	0.1697	0.1011	0.4398	0.2845	0.1061	0.0365

Also in this case, entries in Table 5.6 were used as a selection method for regressor selection of a dynamic model and nine inputs were selected. Gray cells indicate the resultant relevant inputs.

Of course, partial correlation analysis is a truly linear investigation and the designer should take this into account in critically analyzing the results of the investigation.

The same coefficients reported in Table 5.6 are shown in graphical form in Figure 5.25. For the sake of comparison, the same number of model inputs was maintained as with the application of the simple correlation method.

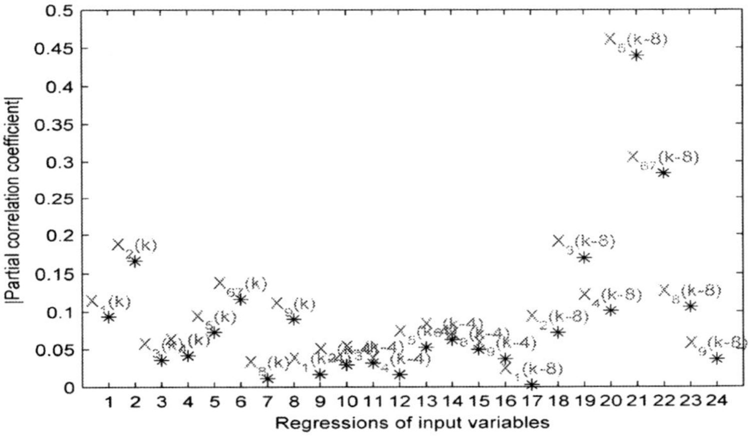

Figure 5.25. Partial correlation coefficients of the debutanizer output variable at time k versus input variables at times k, $k-4$ and $k-8$

It can be observed that the correlation values reported in Table 5.6 are generally lower than the corresponding entries of Table 5.5. This is due to the property of the partial correlation method to isolate the correlation between the two variables under analysis from undesired correlation with other variables.

From the results reported in Tables 5.5 and 5.6, it appears evident that both simple correlation and partial correlation methods give very low coefficient values for any candidate input variable. This gives evidence for the fact that the system could be characterized by strong nonlinearities. Other selection strategies are analyzed in the following subsections.

5.5.3 Mallow's Coefficients with a Linear Model

In Chapter 3 the possibility of using Mallow's C_p coefficient for the selection of influential independent variables was described in some detail. Here the results obtained by computing Mallow's coefficient for data collected from the debutanizer column are analyzed.

As described previously, the C_p coefficient is computed on the basis of the residuals obtained using a given model. In this section; linear models obtained using a classical LMS approach are used.

The computation of Mallow's coefficients is very time consuming and quickly increases with the number of model inputs. For this reason it can represent a suitable selection method when static models are of interest. When regressors must be taken into account, the computational effort could become unaffordable.

Just to give an idea of the nature of information that Mallow's method can give, the results obtained when a static model for the debutanizer column was searched for are reported in Figure 5.26.

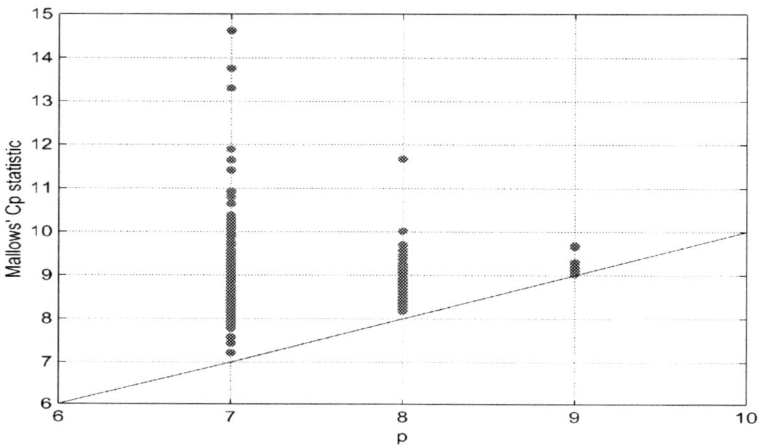

Figure 5.26. Mallow's C_p statistics for models with 7, 8 and 9 inputs, respectively. The suggested model structure corresponds to the point closest to the straight line. In this case a model with nine input variables was suggested

The x-axis reports the class of the combination, *i.e.* in this case models with 7, 8 and 9 inputs were searched for. On the y-axis each '*' represents the value of C_p for a given combination of input variables in the corresponding class. The most adequate model corresponds to the C_p statistics closest to the reported straight line. In the case study considered, a class 9 model is suggested; a static model was not satisfactory and Mallow's statistic was applied for the design of a dynamic model structure. For the purposes of comparison, only 9 input model variables were searched for. This implies that Mallow's C_p coefficient should be estimated on 1307504 subsets of input variables (resulting from class 9 combinations of the 24 inputs). The total computation time was about 30 hours when using Matlab® on a Pentium IV, 2.88 GHz.

The selected inputs are reported in subsection 5.5.6, where the methods considered are compared.

5.5.4 Mallow's Coefficients with a Neural Model

As explained in the previous subsection, Mallow's C_p coefficient depends on model residuals and therefore on the explicit estimation of a particular model with the selected input structure. Though in the previous subsection linear models were taken into account, Mallow's statistic can be computed on wider classes of models.

Here neural models will be considered, in order to take into account the plant nonlinearity.

Such a choice will however further increase the computational effort, because a neural network will now have to be trained for each trial model. Also, a further degree of freedom is introduced, because the optimal number of hidden neurons in the hidden layer (only MLPs with one hidden layer were considered) has to be determined.

In the debutanizer column case study, the computational complexity would be justified by expected improvements. For this reason, a fixed number of neurons was considered after a preliminary trial and error phase.

Moreover, given that a combination of 24 elements of class 9 would have produced an unmanageable number of MLPs to be trained, in this case only 8 inputs were considered; therefore 735471 MLPs with 10 hidden neurons, trained with the Levenberg–Marquardt algorithm, were considered. Even with this simplification, the computation time required, with the same computing structure, would be of the order of weeks, at least when using Matlab® code, and without spending any effort on more suitable programming languages.

Of course, the complexity of the method described with this number of regressors is hardly manageable if a soft sensor designer has to identify a model for a practical application. Here the method was described to give an idea of the complexity the designer is faced with when moving from traditional linear methods to nonlinear extensions.

Selected regressors and model performance are reported in subsection 5.5.6.

5.5.5 PLS-based Methods

In this section, algorithms will be used in a number of ways to help in the selection of model inputs for the debutanizer column.

As a first approach, PLS was applied to the input–output data, where the inputs included corresponding regressors, *i.e.* the same 24 model inputs considered so far. As is known (see Chapter 3), the PLS algorithm projects the input space so that the scores of the measurements provide the maximum information about the output.

The approach is based on the idea of selecting the regressors that contribute to the most important latent variables. From the variance analysis of the input matrix scores, the graph reported in Figure 5.27 was obtained. The first 10 scores out of 24 are reported, along with the explained percentage variance. As can be observed from the figure, latent variables 1 and 3 give the greatest contributions to the explained variance. Looking at the composition of these two latent variables, *i.e.*, the weights of the combinations used to obtain each latent variable, starting from the original 24 model inputs, the nine most relevant unprojected regressors were selected and used to identify a neural-based soft sensor.

In Figures 5.28 and 5.29, the weights used in the linear combination of the 24 original inputs to produce the latent variables 1 and 3 are shown. To select the original input variables that contribute most, the sum of the weights of Figures 5.28 and 5.29 was computed. The results are reported in Figure 5.30 (bars in Figure 5.30 are obtained by adding the corresponding bars reported in Figures 5.28 and 5.29). From Figure 5.30 the nine most relevant original input variables were selected. The number of input regressors was fixed at nine in order to allow comparison of models with the same number of inputs, introduced in previous subsections.

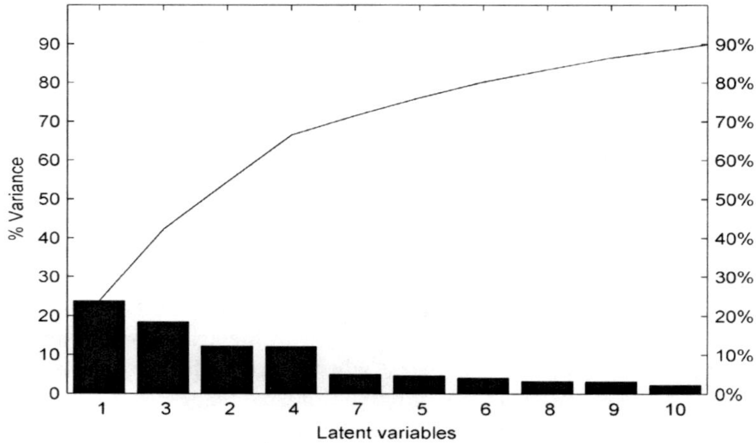

Figure 5.27. Variance analysis of the scores of the input matrix (the first ten scores are reported)

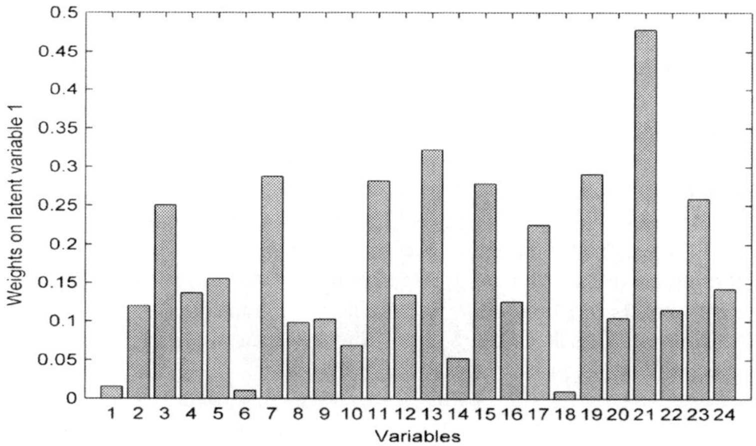

Figure 5.28. Weights used in the linear combination of the 24 original inputs to compute the first latent variable

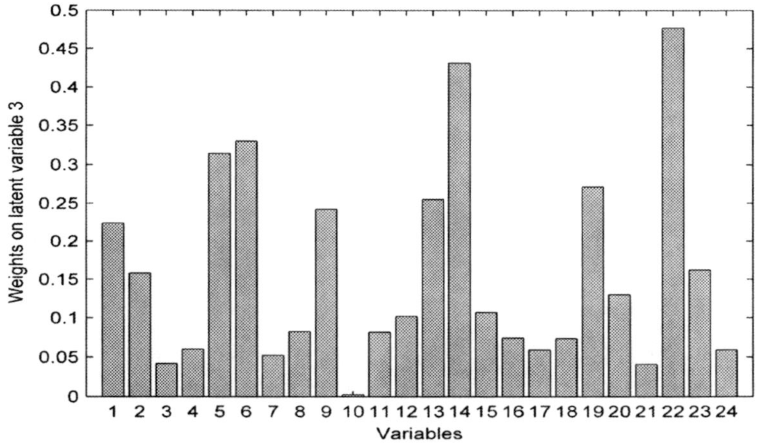

Figure 5.29. Weights used in the linear combination of the 24 original inputs to compute the third latent variable

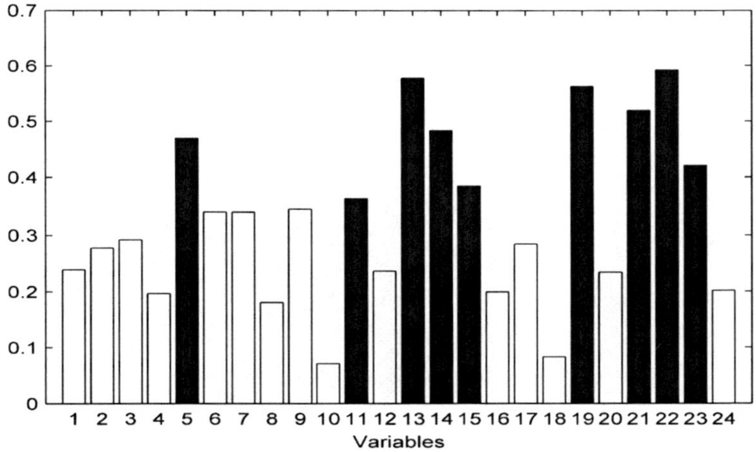

Figure 5.30. Sum of the weights of Figures 5.28 and 5.29. Solid bars indicate the nine most relevant original input variables

The results obtained using the model inputs indicated in Figure 5.30 are reported in the next Section.

The same approach was used to determine a neural-based model for the debutanizer, after application of NPLS, where the nonlinear part of the algorithm is implemented using a set of one input–one output MLP that approximates the nonlinear relationship between each input and output score, as described in Section 5.3. Figures 5.31 to 5.34 show results of the same analysis as that performed in Figures 5.27 to 5.30.

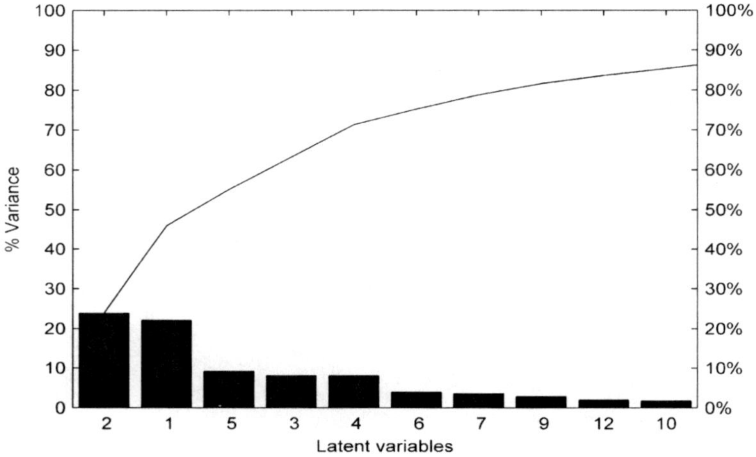

Figure 5.31. Variance analysis of the input matrix scores for the NPLS (the first ten scores are reported)

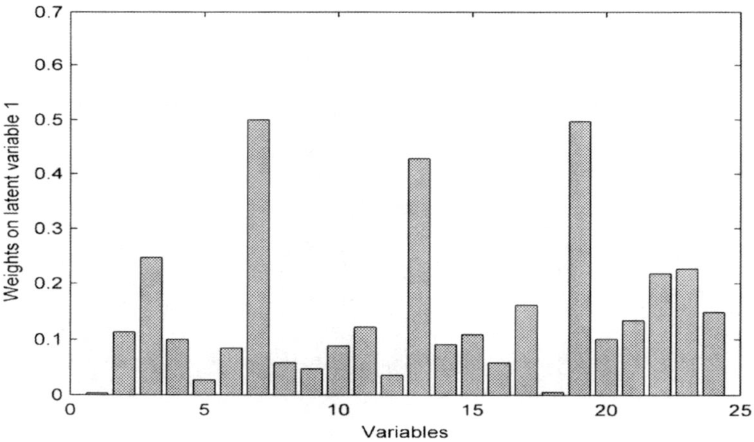

Figure 5.32. Weights used in nonlinear combination of the 24 original inputs to compute the second latent variable

Also in this case, the first two most important latent variables were taken into account. However, this time they proved to be the second and the first latent variables, respectively. After searching for the contribution of the original model input variables to these two latent variables, nine input variables were again taken into account to determine the neural model of the debutanizer column. The results obtained are summarized in the following section.

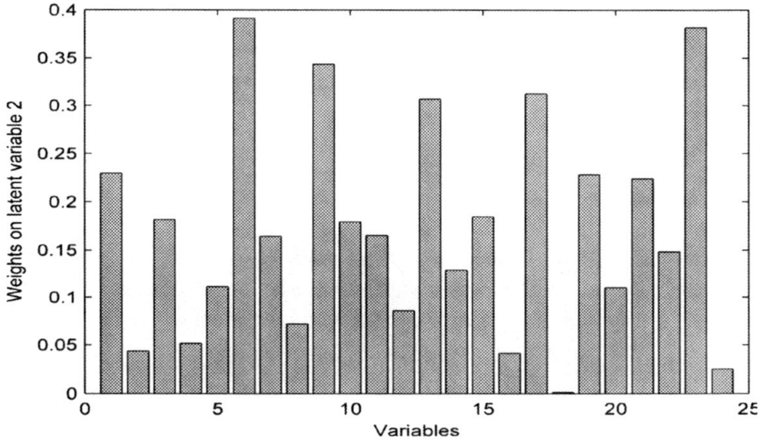

Figure 5.33. Weights used in nonlinear combination of the 24 original inputs to compute the first latent variable

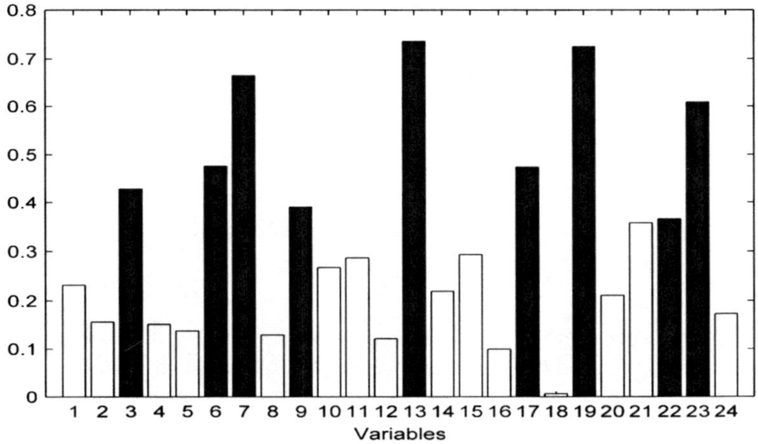

Figure 5.34. Sum of the weights of Figures 5.32 and 5.33. Solid bars indicate the nine most relevant original input variables

Along with the approaches described so far, the classical PLS and NPLS methods reported in the literature were applied to determine the debutanizer model.

For comparison, nine input latent variables were considered both in the case of PLS and in that of its nonlinear extension. It is worth noting that in these last two cases the soft sensor is required to manipulate 24 input variables because dimension reduction of the problem occurs when the internal representation of the problem is performed.

Moreover, in PLS, the input-output representation is truly linear while in NPLS only the internal projection is nonlinear (neural in the case study considered).

The prediction results obtained in these two cases are reported in the following section where, for the sake of completeness, models with 24 input variables will be considered.

5.5.6 Comparison

In this section, variables selected using the various methods described above are reported (Table 5.7) and the corresponding model performance is described. In order to avoid confusion, the following nomenclature is used.

- Model 1: Neural model with inputs selected with simple correlation
- Model 2: Neural model with inputs selected with partial correlation
- Model 3: Neural model with inputs selected with linear Mallow statistics
- Model 4: Neural model with inputs selected with nonlinear Mallow statistics (8 inputs)
- Model 5: Neural model with inputs selected by PLS (unprojected inputs)
- Model 6: Neural model with inputs selected by NPLS (unprojected inputs)
- Model 7: PLS model, with 9 latent variables
- Model 8: NPLS model, with 9 latent variables
- Model 9: PLS model, with 24 latent variables
- Model 10: NPLS model, with 24 latent variables
- Model 11: Neural model with 24 inputs.

In Table 5.7, the regressors selected for Model 1 to Model 6 are reported (other methods are not reported because they use a projected input structure).

It can be observed that the methods described do not agree on the subset of input variables to be used. Moreover, the user has little control over this selection and selected variables depend in a complex way on a number of choices that the designer has to make (*e.g.* the initial number of regressors, selection criteria, number of hidden neurons for MLPs-based models, *etc.*).

Once more, it is clear that the final choice should be made using all available information (including expert knowledge). As an example, for this case study the experts felt more confident with those methods that selected input 5 and its regressors.

Table 5.7. Regressors selected for Model 1 to Model 6

	Model 1	Model 2	Model 3	Model 4	Model 5	Model 6
$x_1(k)$		•	•	•	•	
$x_2(k)$		•	•	•		
$x_3(k)$	•					•
$x_4(k)$						
$x_5(k)$						
$x_{67}(k)$		•				•
$x_8(k)$	•					•
$x_9(k)$		•				
$x_1(k-4)$						
$x_2(k-4)$						
$x_3(k-4)$	•			•	•	
$x_4(k-4)$				•		•
$x_5(k-4)$	•				•	•
$x_{67}(k-4)$			•		•	
$x_8(k-4)$	•				•	
$x_9(k-4)$			•			
$x_1(k-8)$	•			•		•
$x_2(k-8)$			•			
$x_3(k-8)$	•	•	•		•	•
$x_4(k-8)$		•	•	•		
$x_5(k-8)$	•	•	•	•	•	
$x_{67}(k-8)$		•	•	•	•	•
$x_8(k-8)$	•	•	•		•	•
$x_9(k-8)$						

Table 5.8 reports the correlation coefficients obtained testing all the models mentioned above on a set of validation data.

Table 5.8. Correlation coefficient for the debutanizer models

	Correlation coefficient
Model 1	0.8812
Model 2	0.9218
Model 3	0.9245
Model 4	0.9076
Model 5	0.8898
Model 6	0.8745
Model 7	0.7459
Model 8	0.8743
Model 9	0.7502
Model 10	0.9263
Model 11	0.9409

A number of observations can be made when Tables 5.7 and 5.8 are analyzed. Linear models, regardless of the kind of regressor selection criteria and data projection, work less well than nonlinear models (Models 7 and 9).

The introduction of any kind of nonlinearity (*i.e.* a nonlinear model used on data projected by applying PLS, as in the case of Model 5, or linear regression obtained by NPLS, as in the case of Model 8) had a beneficial effect on soft sensor performance.

Further analysis can be performed by looking at the effect of the number of model inputs. The best models were nonlinear structures with the whole set of 24 input variables (Models 10 and 11). However, when models with a reduced number of inputs were of interest, good results were guaranteed by Models 2, 3 and 4. Looking at Table 5.7 it can be observed that these models have some common characteristics. In fact, all of them include input $x_1(k)$ and $x_2(k)$, and consider also the maximum delayed samples of inputs x_4, x_5 and x_{67}.

As a final conclusion, it is observed that for the case study considered, partial correlation analysis and Mallow's statistics were the best methods for reducing the number of model inputs.

Figures 5.35, 5.36 and 5.37 show the results for the most significant cases. Figure 5.35 refers to the worst-performing soft sensor, *i.e.* Model 7. This is not surprising because the model adopted in this case is a linear structure with a reduced number of inputs. This is an interesting consideration since linear models with input selection based on PLS are widely used in industrial applications.

Figure 5.36 reports the results obtained by using Model 3, because this is the best model among those with a reduced set of inputs.

Finally, Figure 5.37 reports tests performed on Model 11 since this is the best-performing model.

In all cases considered, both time plots of the model output and 4-plot analysis of the residuals are reported.

From the analysis of the reported figures it can be observed that, in accordance with the correlation coefficients reported in Table 5.8, the higher the correlation coefficient, the closer the simulated data matches the acquired data.

Also, the corresponding 4-plot analysis confirms that for soft sensors with better performance, the corresponding residuals are closest to being uncorrelated.

The model validation reported so far is not exhaustive, since in this chapter attention was focused on model structure selection.

A deeper description of methods for model validation is reported in the next chapter.

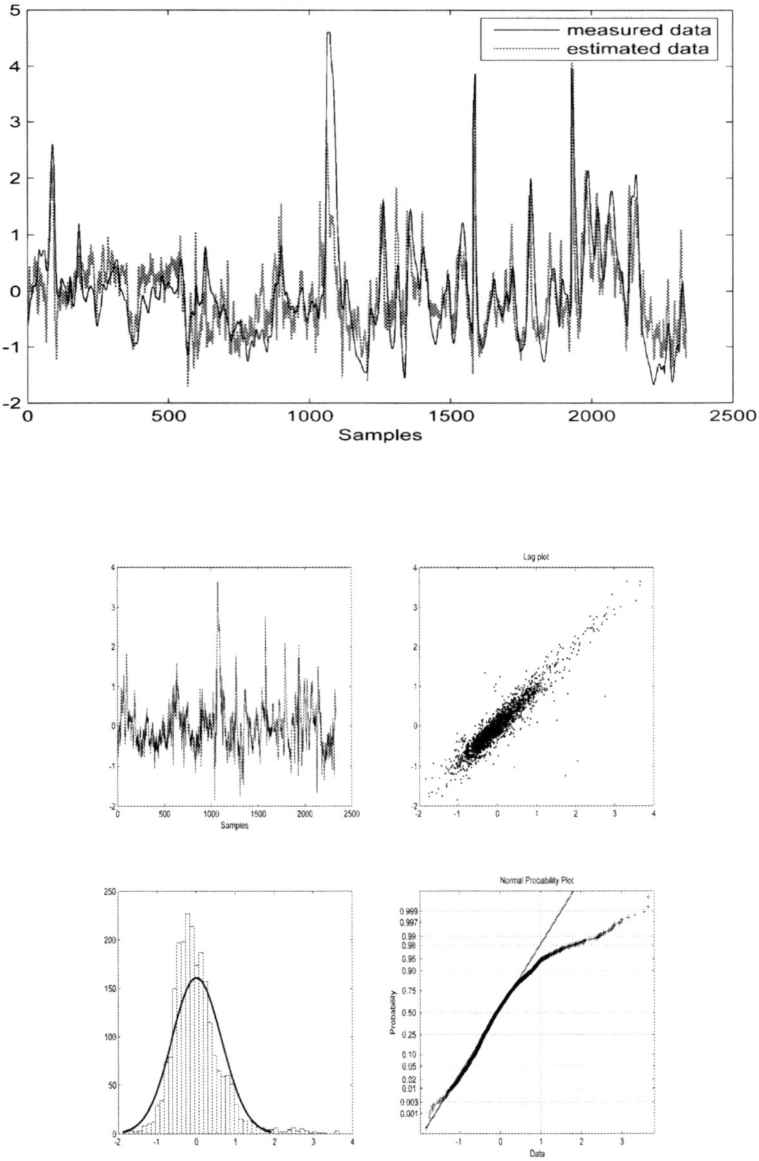

Figure 5.35. Comparison between measured data and estimations obtained with Model 7 and 4-plot analysis of corresponding residuals

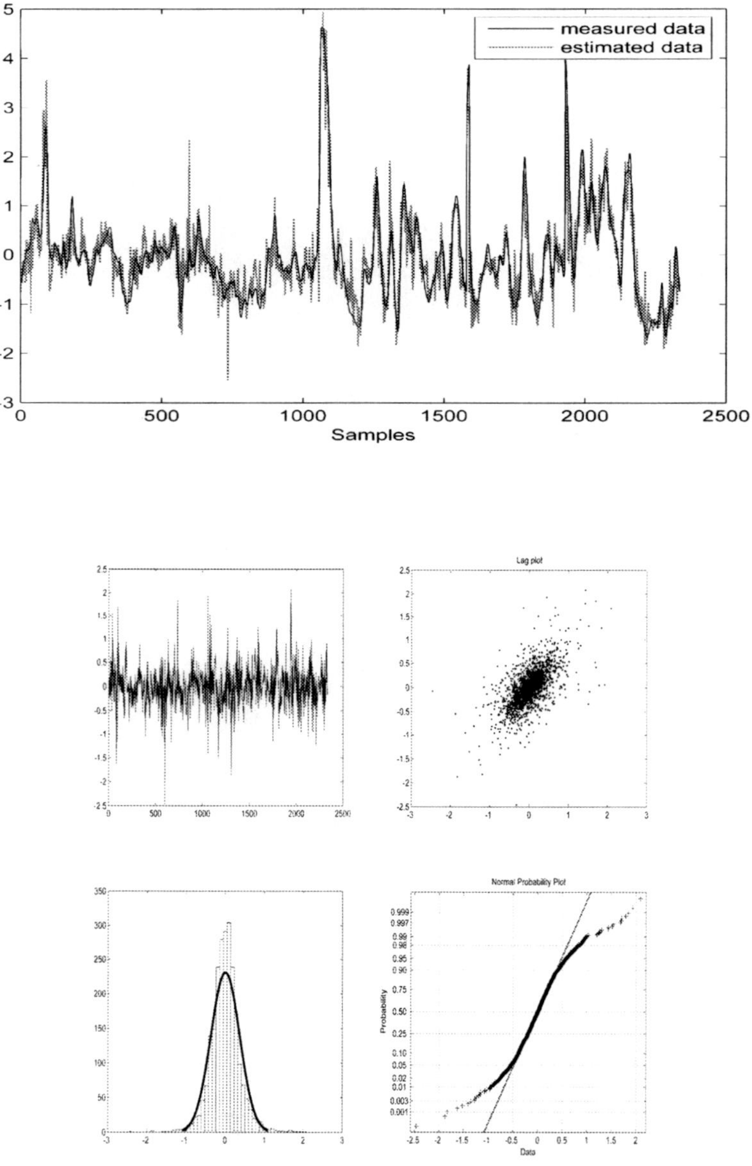

Figure 5.36. Comparison between measured data and estimations obtained with Model 3 and 4-plot analysis of corresponding residuals

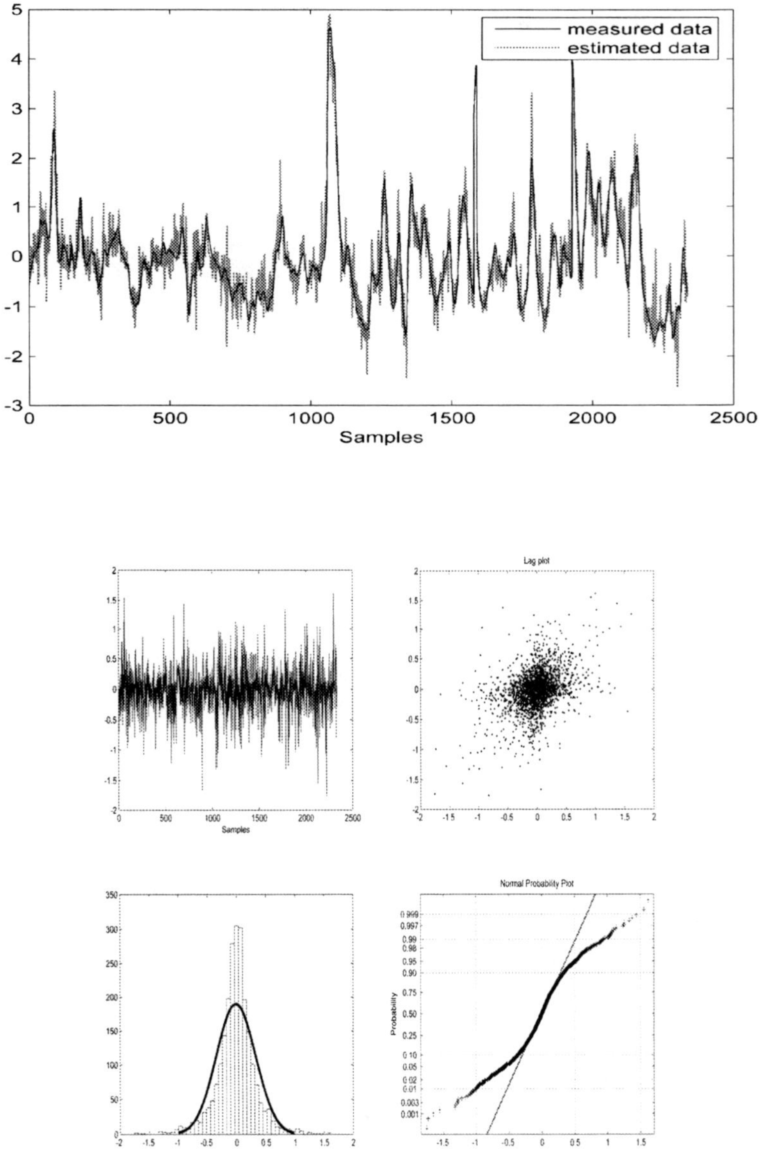

Figure 5.37. Comparison between measured data and estimations obtained with Model 11 and 4-plot analysis of corresponding residuals

5.6 Conclusions

Model structure selection is the core of any data-driven system modeling. The very first choice is between static and dynamic models. No real system can be considered strictly static. Nevertheless, if the input signal frequency content is much lower than time scales involved in energy exchange phenomena occurring in the system, a static model can be an accurate approximation. From a practical point of view, static models can be an adequate choice if system transients are not relevant to the application. In the case of static models, PCA and PLS are valid tools to further simplify the modeling task, avoiding the negative effects of data collinearity and reducing the problem dimensionality.

The case of dynamic modeling is a much more complex task because the designer needs to select the model structure, the input variables and the regressors. The objective of the soft sensor will drive the designer in the choice between auto-regressive or moving-average models, as discussed in Section 2.2.

The problem of a dynamic model structure selection can be a very time consuming task to be solved, especially in the case of MIMO systems. This requires exploiting any physical insight on the plant.

A starting point of any design procedure should be to consider a linear structure. This allows running a large number of trials using system identification software tools, largely available on the market. These tools mainly use LMS based algorithms to design the system parameters and correlation analysis to analyze the performance of model candidates. Also in the case of dynamic system PCA and PLS can be used to reduce the model complexity. More sophisticated techniques, such as partial correlation method, Mallow's statistics, and Lipschitz quotients can be helpful to select model inputs and their regressors.

Linear models can give good performance when the model is required to represent the system dynamic close to a fixed working point. If this is not the case a great improvement can be obtained using nonlinear models. This introduces further degrees of freedom in the modeling task, represented by the nonlinearity structure. Unless some physical considerations suggest for a particular nonlinearity, "universal approximators" are widely used and neural networks are by far the most popular and effective choice.

In the cases when linguistic system knowledge is available, fuzzy approximators can be successfully used. In such a case the model obtained can have the interesting property to give a good understanding of the process.

6

Model Validation

6.1 Introduction

In this chapter the problem of model validation, theoretically described in Chapter 3, will be considered with reference to a case study. In that chapter it was outlined that the problem of model validation mainly consists in verifying that model residuals are not correlated with model inputs and that their autocorrelation function is an impulse function.

The procedures that should be followed in the case of linear systems are quite well established and a huge amount of literature exists on this topic (Ljung, 1999). In particular, the usual statistical approach for linear model validation mainly consists in evaluating the characteristics of the autocorrelation function of the residuals and the cross-correlation function between the residuals and the inputs.

When dealing with nonlinear systems, the checking of residual properties is more cumbersome; however, a number of tests designed to detect whether the residual is unpredictable from all linear and nonlinear combinations of past inputs and outputs is reported in Billings, Jamaluddin and Chen (1992), Billings and Voon (1991) and Mendes and Billings (2001)). These tests were introduced when considering the case of analytic nonlinear models, anyway they are generally used for neural network based NARMAX models. In fact, while a theoretical analysis of this aspect does not exist, experimental evidence of their usefulness in the case of neural models is amply available. In practice, model validation requires that normalized correlation functions between couples of sequences ψ_i and ψ_j be estimated. The sampled correlation function is

$$\hat{\phi}_{\psi_i\psi_j}(\tau) = \frac{\sum_{t=1}^{N-\tau} \psi_i(t)\,\psi_j(t+\tau)}{\sqrt{\sum_{t=1}^{N} \psi_i^2(t) \sum_{t=1}^{N} \psi_j^2(t)}} \tag{6.1}$$

The following conditions need to be tested, for each input variable:

$$\phi_{\varepsilon\varepsilon}(\tau) = E[\varepsilon(t-\tau)\varepsilon(\tau)] = \delta(\tau) \tag{6.2}$$

$$\phi_{u\varepsilon}(\tau) = E[u(t-\tau)\varepsilon(\tau)] = 0 \quad \forall\tau \tag{6.3}$$

$$\phi_{u^{2'}\varepsilon}(\tau) = E[(u^2(t-\tau) - u_m^2(t))\,\varepsilon(t)] = 0 \quad \forall\tau \tag{6.4}$$

$$\phi_{u^{2'}\varepsilon^2}(\tau) = E[(u^2(t-\tau) - u_m^2(t))\varepsilon^2(t)] = 0 \quad \forall\tau \tag{6.5}$$

$$\phi_{\varepsilon(\varepsilon u)}(\tau) = E[\varepsilon(t)\,\varepsilon(t-1-\tau)\,u(t-1-\tau)] = 0 \quad \tau \geq 0 \tag{6.6}$$

where $E[\cdot]$ is the mathematical expectation, $\varepsilon(t)$ is the model residual, $u(t)$ is the generic input variable, and the subscript m denotes the time average. The normalization adopted in Equation 6.1 guarantees that correlation introduced functions are in the interval [−1, 1].

In practical applications none of the reported correlation functions is exactly zero for any of the lags considered. Instead, they are traced with a corresponding confidence band. The 95% confidence band is generally used and is $1.96/\sqrt{N}$, where N is the number of samples considered.

The presence of values of correlation functions lying significantly outside the confidence band for any time lag suggests that it is advisable to consider the corresponding lagged input in the model structure.

In the next section, a fairly complex case study will be analyzed in detail and the complete validation of two different neural network based models will be reported in order to select the more suitable soft sensor. Such a choice will be the result of comprehensive analysis of a number of quantitative and qualitative evaluations, mainly regarding time plots and residual analysis, on the one hand, and computational complexity on the other.

6.2 The Debutanizer Column

The case study considered in this section is the real-time prediction of the butane concentration (C4) in the bottom flow of a debutanizer column, on the basis of a set of available measurements. This case study was considered in Chapter 5, where a number of strategies to select model inputs and regressors in order to design an NMA model, were analyzed.

In the present chapter, a different strategy has been adopted to design a NARX model in which input variable selection has been performed with the help of an expert with a subsequent attempt to refine regressor selection using correlation analysis. Of course, the two approaches mentiond should not be considered as exclusive and can be used in conjunction with each other.

The soft sensors have been designed to overcome the problem of the delay introduced by the gas chromatograph because of its location. In fact, the C4 content in the debutanizer bottoms is not detected in the bottom flow, but on the overheads of a deisopentanizer column where it measures the content of C4 in the isopentane flow (iC5) to stock, as reported in the Appendix.

The C4 content in iC5 depends exclusively on the debutanizer operating conditions: the C4 detected in the iC5 flow can be assumed to be that coming from the debutanizer bottom.

The measuring device has a measuring cycle lasting 15 min. Because of the analyzer location, concentration values are obtained with a long delay, which has been estimated to be about 30 min. This means that information about C4 is available with a delay of about 45 min and, since this was considered not suitable for control purposes, a soft sensor was required.

Despite the presence of the delay, acquired data about system output variables are available for use in the plant model. In addition, a very accurate model is required. In view of these two considerations, a NARX model is considered the most adequate model structure.

Two different structures can be adopted. The first approach is based on the use of a cascaded structure comprising one-step-ahead predictors of the plant output $\hat{y}(k)$ in the form:

$$\hat{y}(k) = f(\bar{u}(k),\ldots,\bar{u}(k-m), \hat{y}(k-1),\ldots, \hat{y}(k-3), y(k-4),\ldots, y(k-n)) \quad (6.7)$$

where the estimated output samples $\hat{y}(k-1)$, $\hat{y}(k-2)$ and $\hat{y}(k-3)$ are computed using Equation 6.7 at previous time steps, and older samples are available in the plant database.

The second approach uses a model in the form:

$$\hat{y}(k) = f(\bar{u}(k),\ldots,\bar{u}(k-m), y(k-4), y(k-5),\ldots, y(k-n)) \quad (6.8)$$

where the sampling time used in this application is $T_s = 12$ min, so that $k-4$ corresponds to a delay of 48 min and the regressor y(k−4) considered for the output is the newest measured sample available in the plant database (*i.e.* it does not reflect a system delay).

In both cases, the input regressor vector $\bar{u}$ has been chosen using a trial and error approach, guided by the plant experts. The results obtained with these two approaches are reported in the next two sections. Furthermore, in Section 6.4 a refinement of the one-step-ahead soft sensor, suggested by the correlation analysis is reported.

6.3 The Cascaded Structure for the Soft Sensor

The cascaded structure is made by using four one-step-ahead predictors organized as reported in Figure 6.1.

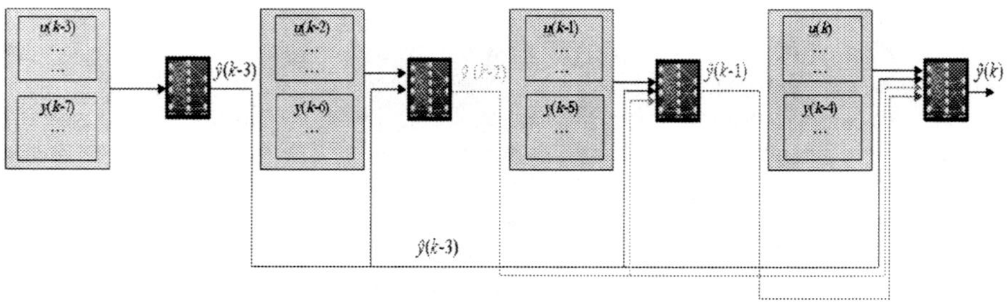

Figure 6.1. Block scheme of the cascaded neural network based soft sensor. Inputs to each network are either from previous blocks or from the plant database

The blocks are realized using the same neural network, which at each step receives different inputs, either from the previous block or from the plant database. The neural block in Figure 6.1 implements the following model

$$\hat{y}(k) = f\Big(u_1(k), u_2(k), u_3(k), u_4(k), u_5(k), u_5(k-1), u_5(k-2), u_5(k-3), \left(\frac{u_6(k)+u_7(k)}{2}\right), \hat{y}(k-1), \hat{y}(k-2), \hat{y}(k-3), y(k-4)\Big) \tag{6.9}$$

and is an MLP with 12 hidden neurons.

The performance of the soft sensor in Figure 6.1 computed on a set of validation data is reported in Figures 6.2 and 6.3. In Figure 6.2, acquired values of C4 are compared with their real-time estimations. Figure 6.3 shows the 4-plot analysis of the corresponding model residuals.

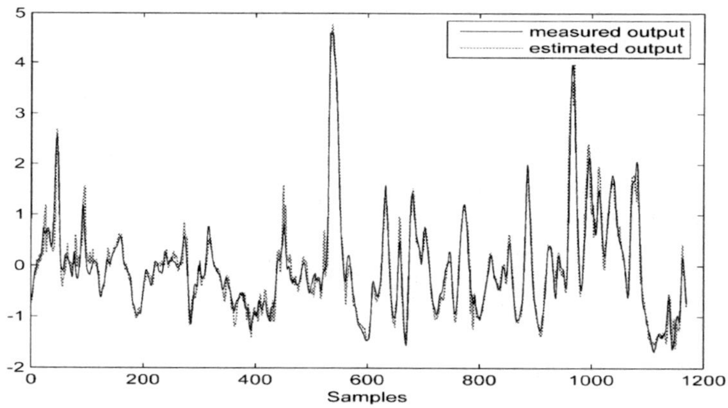

Figure 6.2. Comparison of the real-time estimation of C4 concentration in the bottom flow of the debutanizer column with acquired data

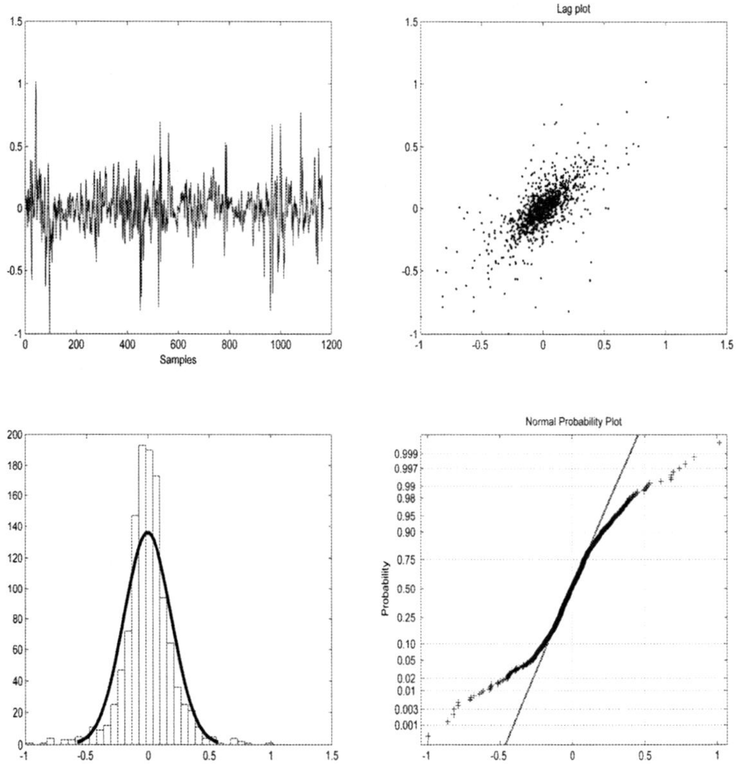

Figure 6.3. 4-plot analysis of the residuals of real-time estimation of C4 concentration in the bottom flow of the debutanizer column

The correlation coefficients for the output estimations of each block in Figure 6.1 are reported in Table 6.1.

Table 6.1. Correlation coefficient values for output estimations in the cascaded structure shown in Figure 6.1

	Corr. coefficient
Step 1	0.99796
Step 2	0.99441
Step 3	0.98604
Step 4	0.96592

A complete analysis of the estimated correlation functions expressed by Equations 6.1 to 6.6 is reported in the following paragraphs.

It is worth noting that, on the basis of the number of model inputs, each of the Equations 6.3 to 6.6 requires the estimation of 13 different correlation functions. If the model structure reported in Equation 6.9 is considered carefully it is evident that independent inputs reduce to seven, while other variables are either delayed samples of the input variables or of the model output. For this reason, some of the estimated correlation functions will be no more than translated copies. This is exactly the case for the correlation between the model residual and delayed copies of the inputs. Such a conclusion does not hold totally true for the model output, because the recursive structure of the soft sensor introduces slight differences in the various delayed estimations of the system output (see the comments on Table 6.1 reported above). In fact, if the four correlation functions between the model residuals and the output estimation delayed samples are examined, a progressive change in their shape can be observed. The autocorrelation function of Equation 6.1 is reported in Figure 6.4, while correlation functions between each input and the model residual are reported in Figures 6.5 to 6.9. Note that correlation functions reported in Figure 6.7 correspond to regressors of the same input variable u_5. The same consideration is valid for Figure 6.9 where regressors of the model output are considered.

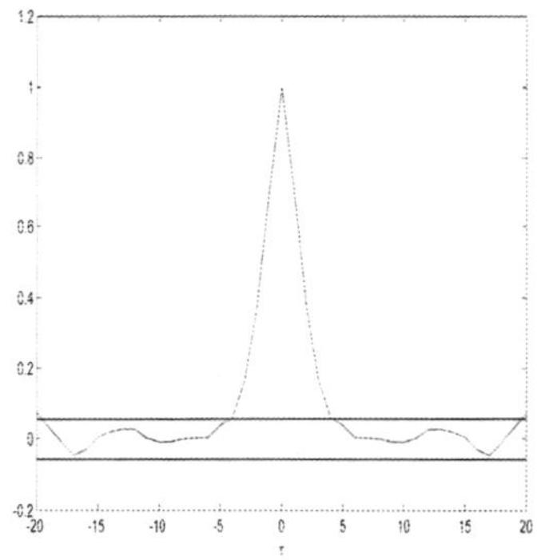

Figure 6.4. Autocorrelation function of Equation 6.2 for the model residual

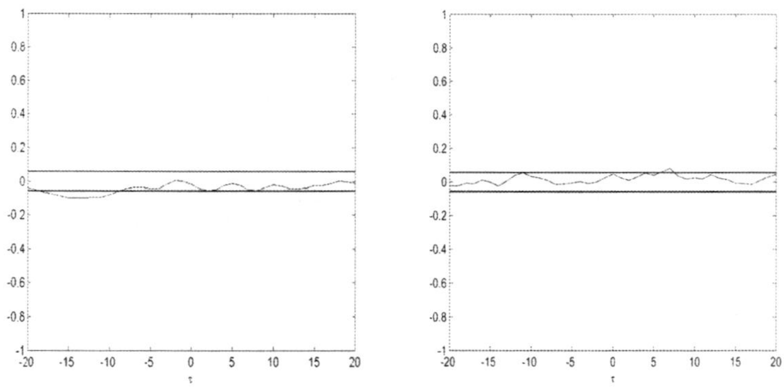

Figure 6.5. Correlation functions corresponding to Equation 6.3, inputs u_1 and u_2

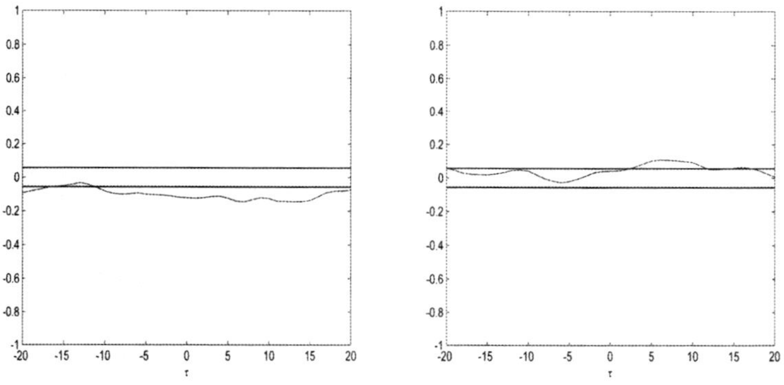

Figure 6.6. Correlation functions corresponding to Equation 6.3, inputs u_3 and u_4

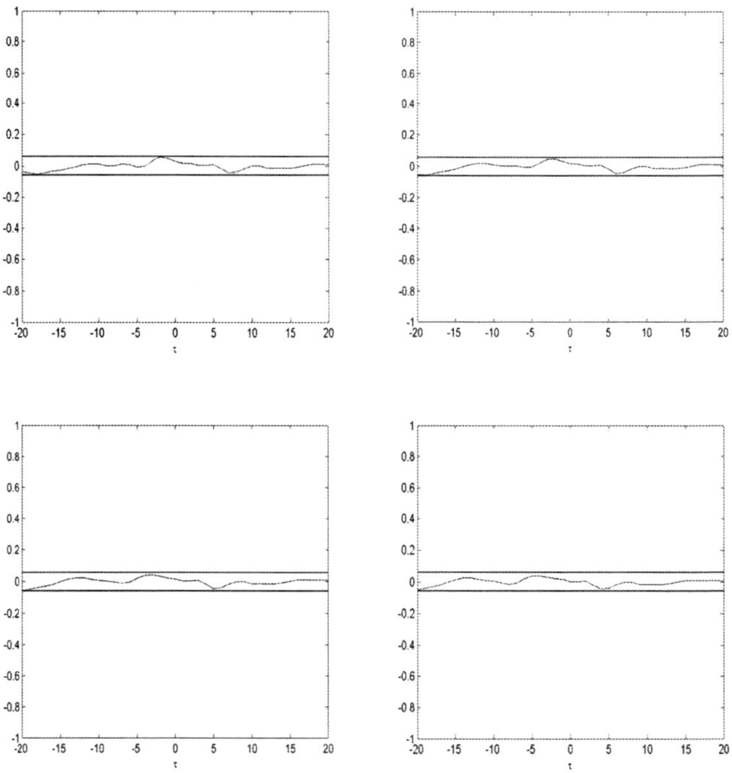

Figure 6.7. Correlation functions corresponding to Equation 6.3, inputs u_5 and its delayed replicas

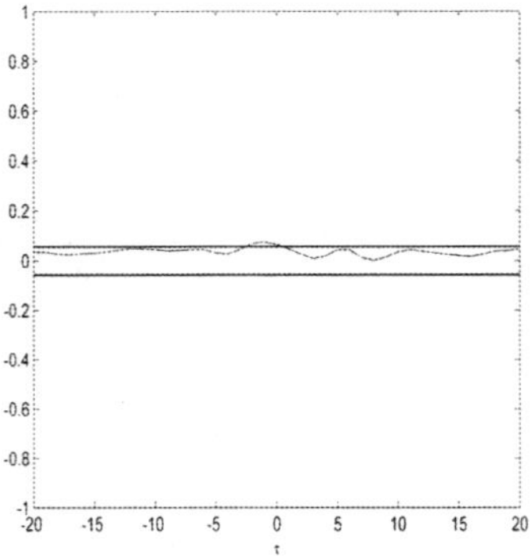

Figure 6.8. Correlation function in Equation 6.3, input $(u_6 + u_7)/2$

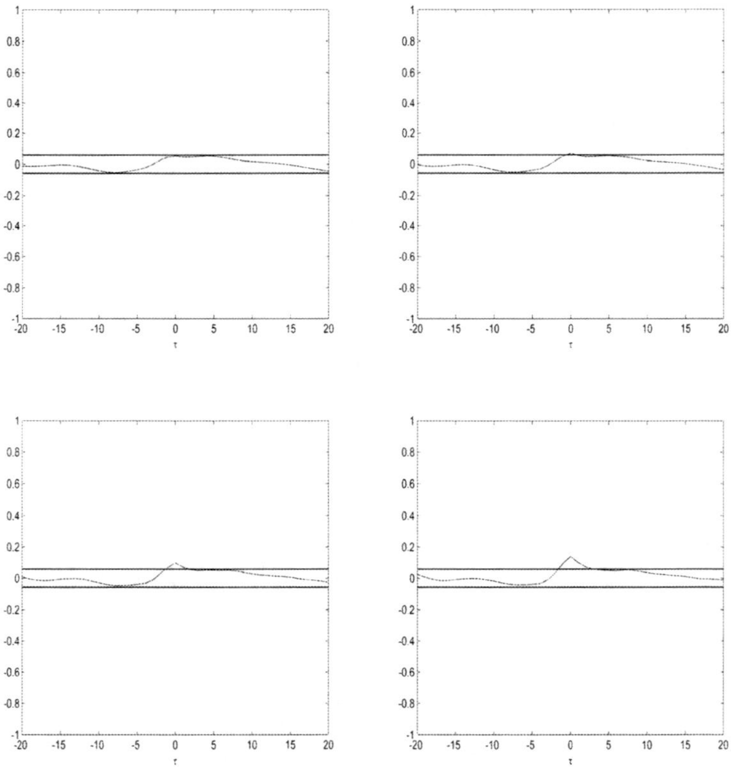

Figure 6.9. Correlation functions corresponding to Equation 6.3, output and its regressors

In what follows, results corresponding to Equations 6.4, 6.5 and 6.6 are reported. Only independent model inputs are considered.

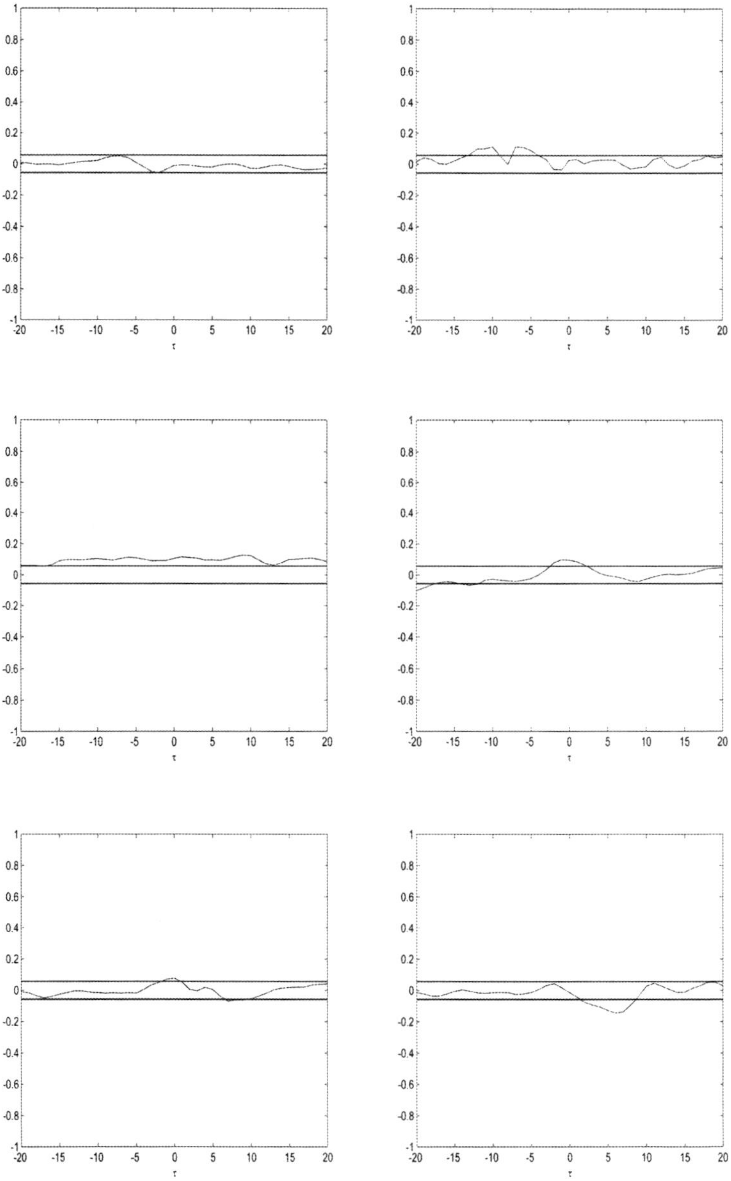

Figure 6.10. Correlation functions for Equation 6.4, independent model inputs

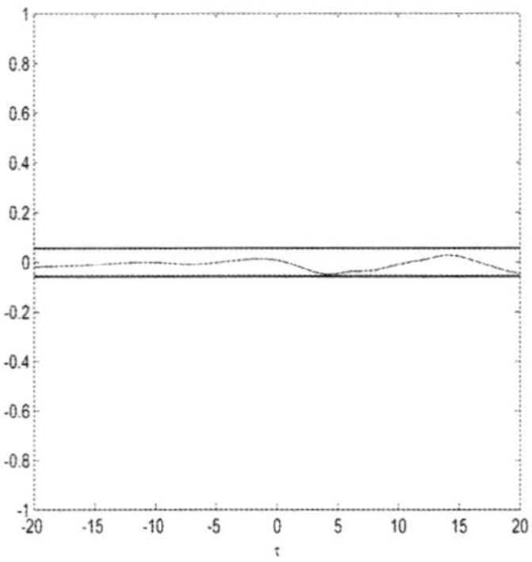

Figure 6.11. Correlation function for Equation 6.4 for the plant output

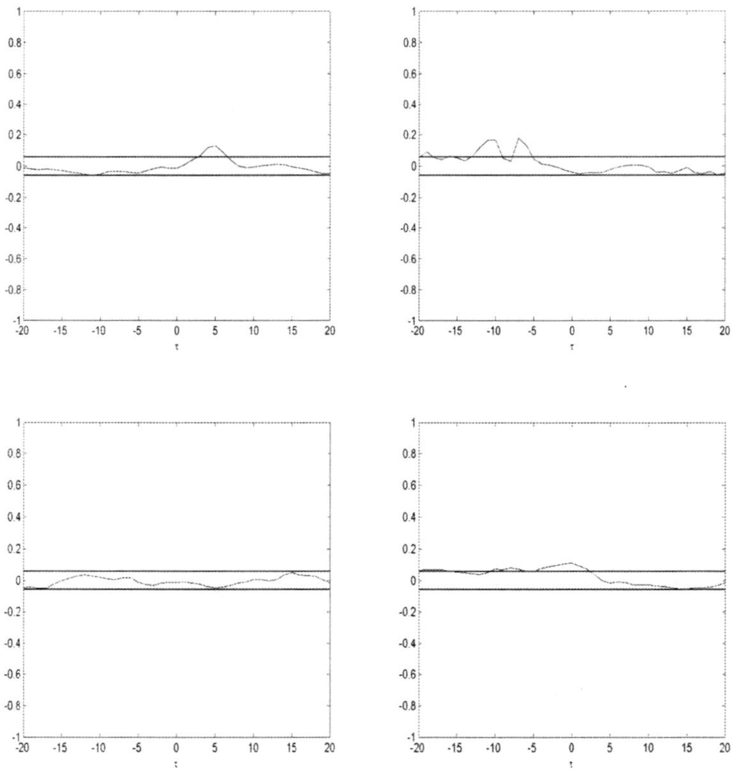

Figure 6.12. Correlation functions for Equation 6.5, plant inputs u_1 to u_4

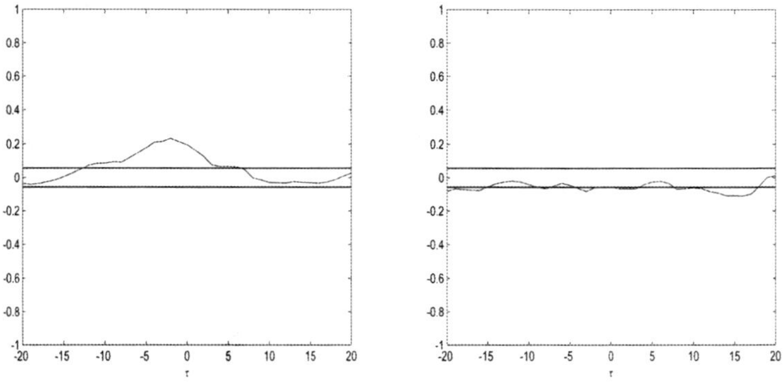

Figure 6.13. Correlation functions for Equation 6.5, inputs u_5 and $(u_6 + u_7)/2$

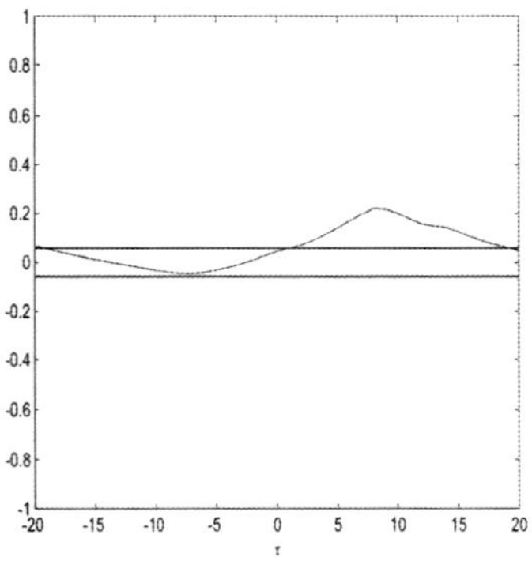

Figure 6.14. Correlation function for Equation 6.5 for the plant output

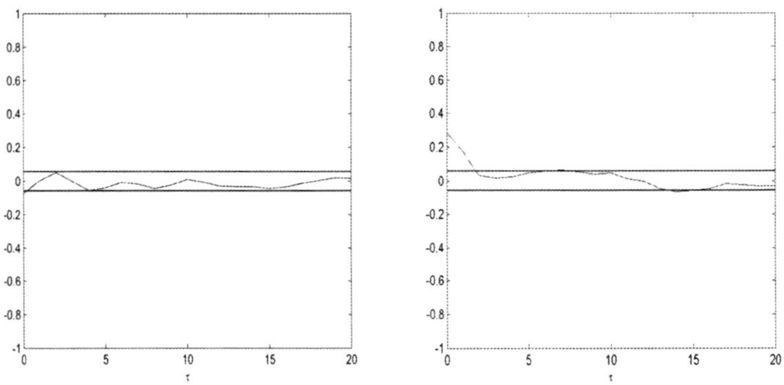

Figure 6.15. Correlation functions for Equation 6.6, plant inputs u_1 and u_2

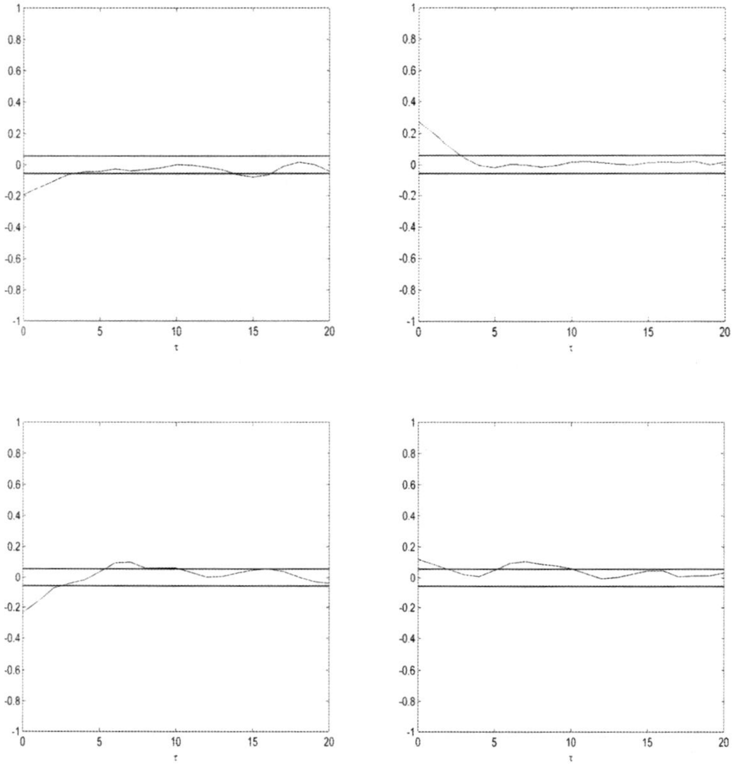

Figure 6.16. Correlation functions for Equation 6.6, inputs from u_3 to $(u_6 + u_7)/2$

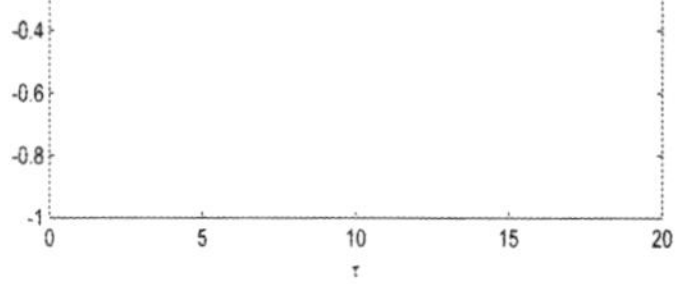

Figure 6.17. Correlation functions for Equation 6.6, plant output

Though some of the reported correlation functions do not lie in the corresponding 95% confidence band, the model simulation capability looks quite satisfactory, as can be observed in Figure 6.2, reported at the beginning of this subsection.

Moreover, any attempt to improve the model performance on the basis of correlation functions analysis produced no significant improvement. Nevertheless, in the following subsection, the results obtained by applying the second structure are reported and some final comments drawn.

6.4 The One-step-ahead Predictor Soft Sensor

In this second case a one-step-ahead model is directly designed as the core of the soft sensor, using already available output regressors, *i.e.* old enough for them to have been already measured by the analyzer and stored in the corresponding plant database.

Based on this consideration, the structure of the output regressors in the model was adapted to fit the new constraints, while the structure for the input regressors was maintained the same as in Equation 6.9 of the previous subsection. The neural model now has the following structure:

$$\begin{aligned}\hat{y}(k) = f(\ & u_1(k), u_2(k), u_3(k), u_4(k), u_5(k), u_5(k-1), u_5(k-2), u_5(k-3), \\ & \left(\frac{u_6(k)+u_7(k)}{2}\right), y(k-4), y(k-5), y(k-6)\ )\end{aligned} \tag{6.10}$$

Also in this case, the model was implemented using an MLP with 12 hidden neurons. The model guaranteed a correlation coefficient of 0.982.

The graphics relative to the model described by Equation 6.10, computed using the same set of validation data as that considered in Subsection 6.2.1 for the cascaded structure, are reported in Figures 6.18 and 6.19. In Figure 6.18, acquired values of C4 are compared with their real-time estimations, while the 4-plot analysis of model residuals is shown in Figure 6.19.

Here again we report, in Figures 6.20 to 6.31, the correlation analysis of the model residuals on the basis of Equations 6.2 to 6.6. Only independent model inputs are considered.

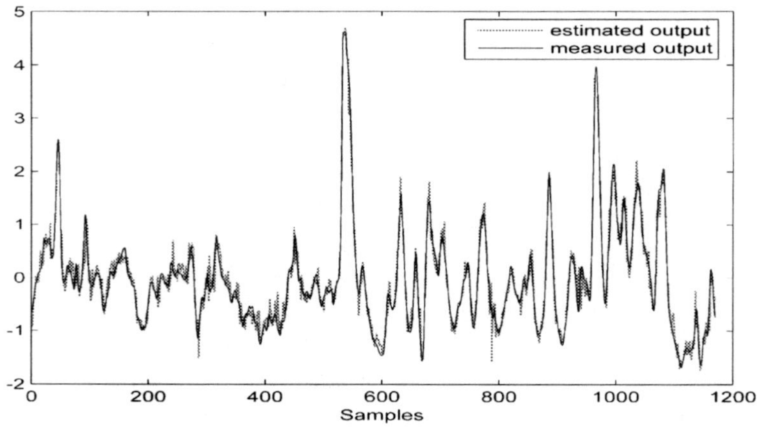

Figure 6.18. Comparison of measured and real-time estimation of C4 concentration in the bottom flow of the debutanizer column obtained using the one-step-ahead model

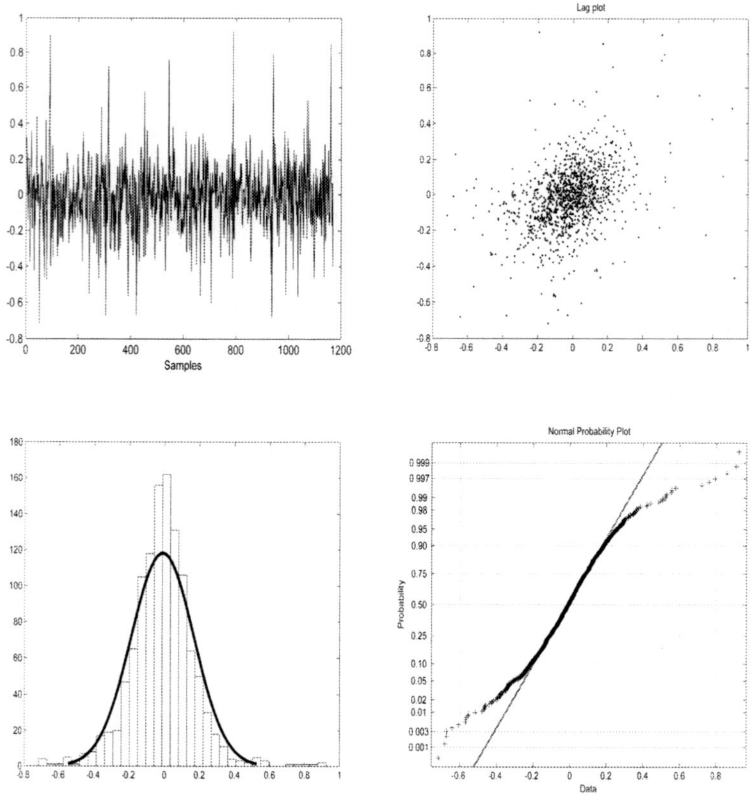

Figure 6.19. 4-plot analysis of the one-step-ahead model residuals

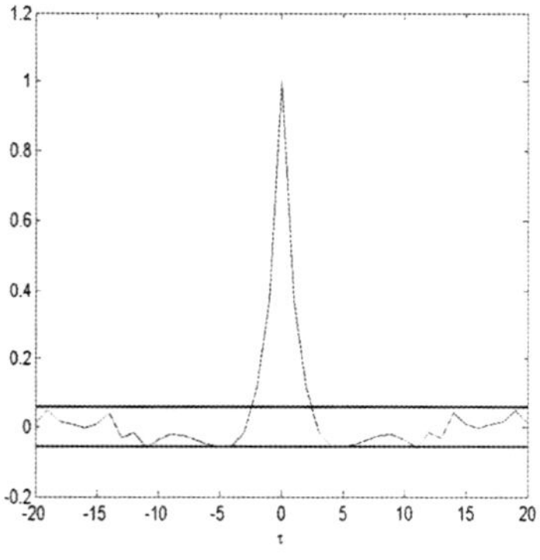

Figure 6.20. Autocorrelation function of Equation 6.2 for the model residuals

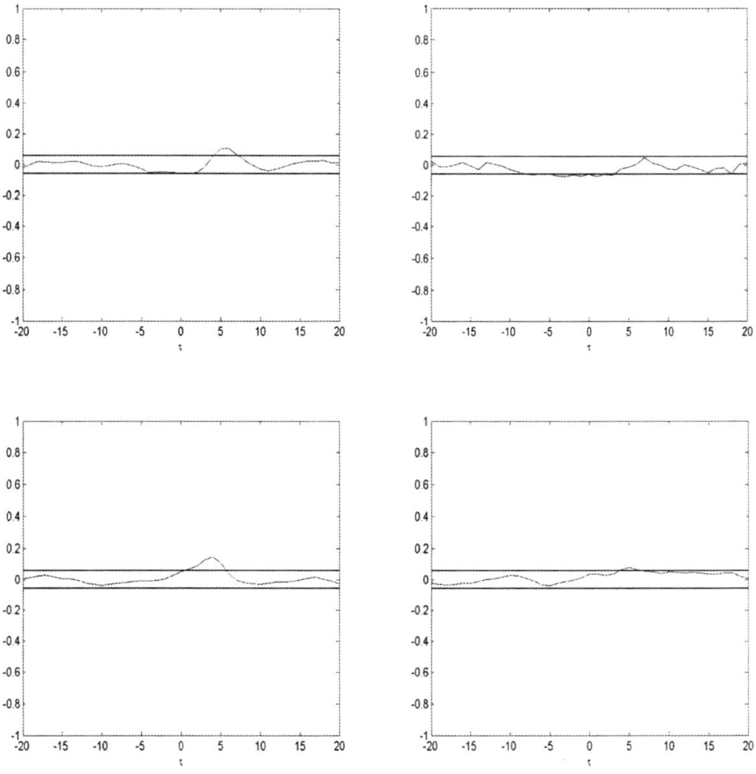

Figure 6.21. Correlation functions of Equation 6.3. Model inputs from u_1 to u_4

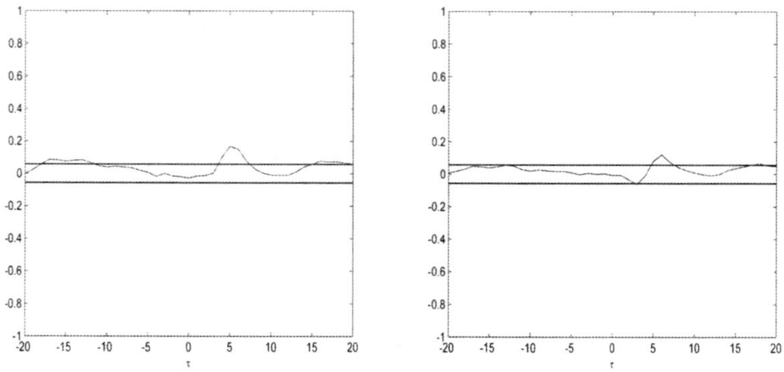

Figure 6.22. Correlation functions of Equation 6.3. Model inputs u_5 and $(u_6+u_7)/2$

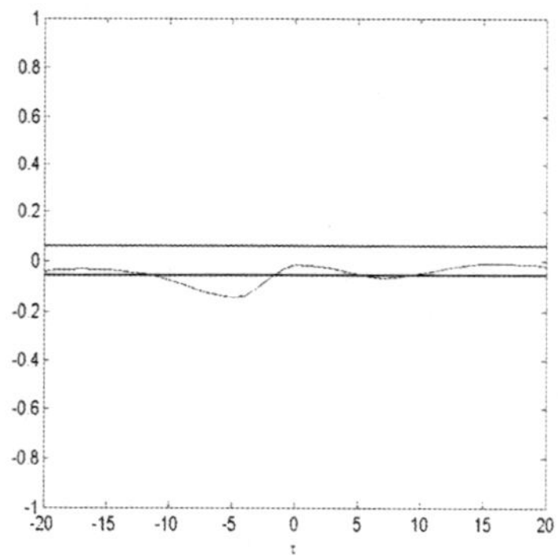

Figure 6.23. Correlation function of Equation 6.3. Plant output

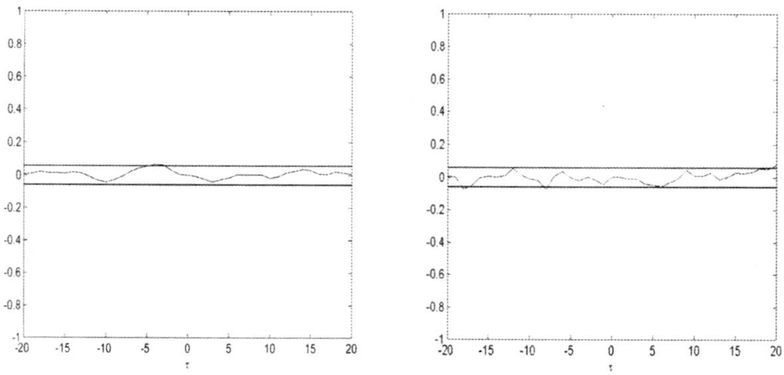

Figure 6.24. Correlation functions for Equation 6.4. Model inputs u_1 and u_2

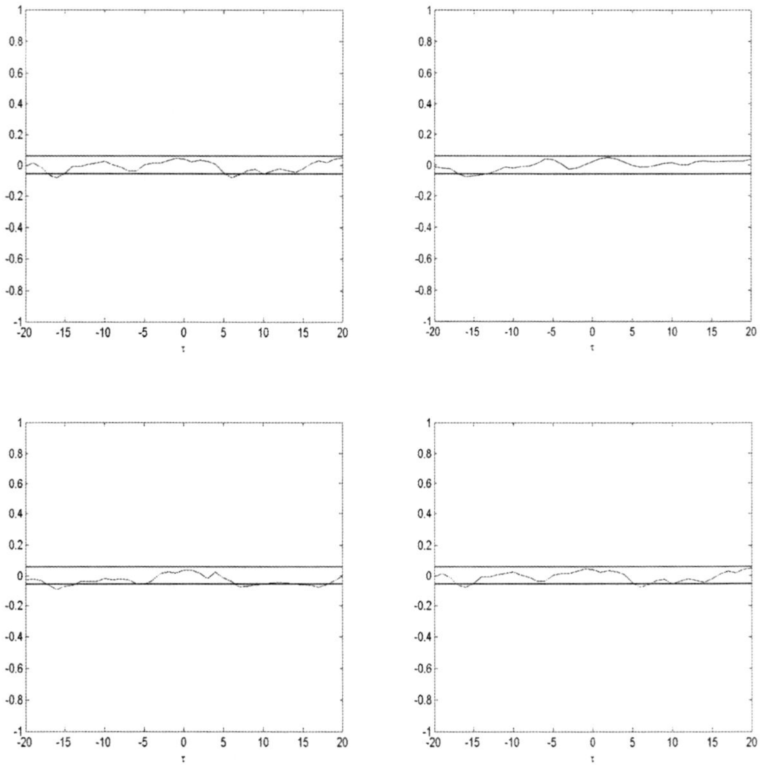

Figure 6.25. Correlation functions for Equation 6.4. Inputs u_3 to u_5 and input $(u_6+u_7)/2$

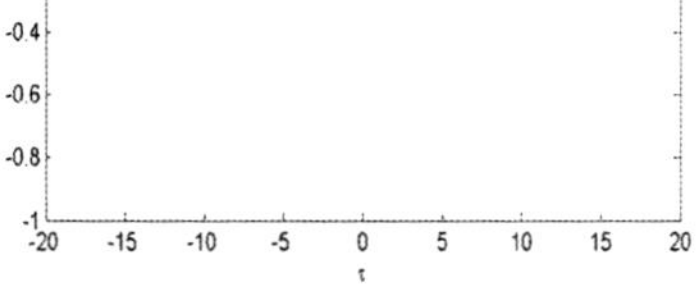

Figure 6.26. Correlation function for Equation 6.4 for the plant output

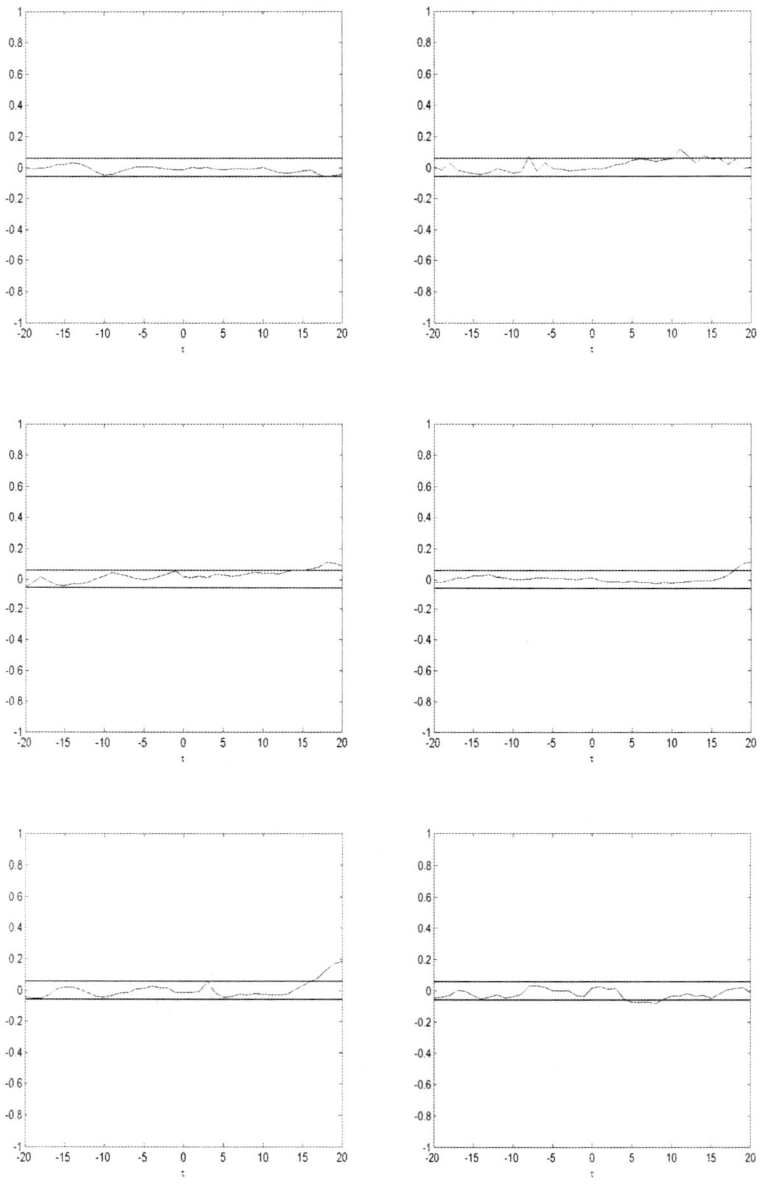

Figure 6.27. Correlation functions for Equation 6.5. Inputs from u_1 to $(u_6+u_7)/2$

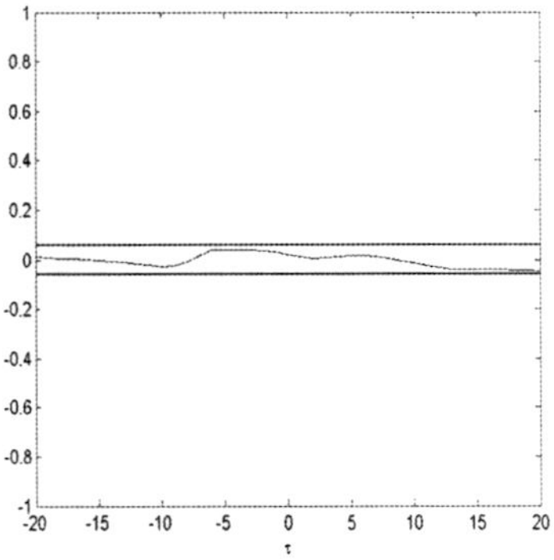

Figure 6.28. Correlation function for Equation 6.5 for the plant output

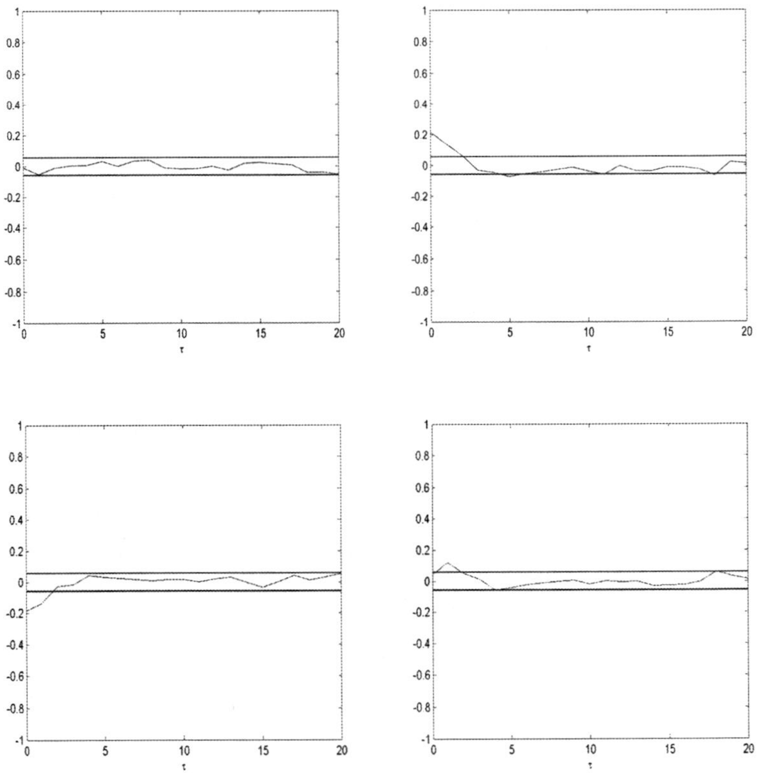

Figure 6.29. Correlation functions for Equation 6.6. Inputs u_1 to u_4

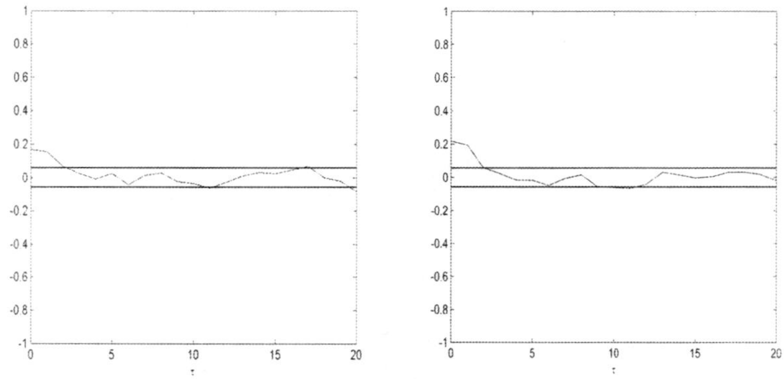

Figure 6.30. Correlation functions for Equation 6.6. Inputs u_5 and $(u_6+u_7)/2$

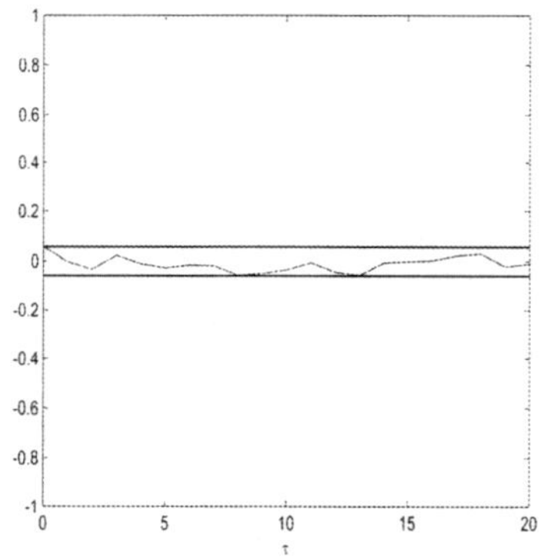

Figure 6.31. Correlation function for Equation 6.6 for the plant output

The analysis of the correlation graphs shows that a significant correlation can still be hypothesized between inputs and model residuals.

A new model structure can be tested by adding regressors corresponding to model input samples that produce peaks in the correlation functions outside the reported confidence band. In fact, this is evidence of a residual significant correlation between the model input considered and the corresponding estimated output.

In the following subsection, the results of this attempt are reported.

6.4.1 Refinement of the One-step-ahead Soft Sensor

On the basis of the visual inspection of correlation graphs reported in this section, the model structure was enlarged, adding a number of model input regressors when values of the correlation functions outside the confidence interval were observed.

The model structure was modified as follows

$$\begin{aligned}\hat{y}(k) = f(\, & u_1(k), u_1(k-5), u_2(k), u_3(k), u_3(k-5), u_4(k), \\ & u_5(k), u_5(k-1), u_5(k-2), u_5(k-3), u_5(k-4), \\ & \left(\frac{u_6(k)+u_7(k)}{2}\right), \left(\frac{u_6(k-5)+u_7(k-5)}{2}\right), \\ & y(k-4), y(k-5), y(k-6)\,)\end{aligned} \tag{6.11}$$

Once again, the model that performed best was an MLP with 12 hidden neurons. The model guaranteed a correlation coefficient of 0.985. This means that the enlarged model guaranteed only a slight improvement in correlation coefficient.

The graphics relative to the modeling capability obtained using Equation 6.11 are reported in Figures 6.32 and 6.33.

In Figure 6.32 the estimations obtained from the model are compared with acquired data, while Figure 6.33 shows the results of the 4-plot analysis of the model residuals.

In Figures 6.34 to 6.44, the correlation analysis of the model residuals are reported on the basis of Equations 6.2 to 6.6.

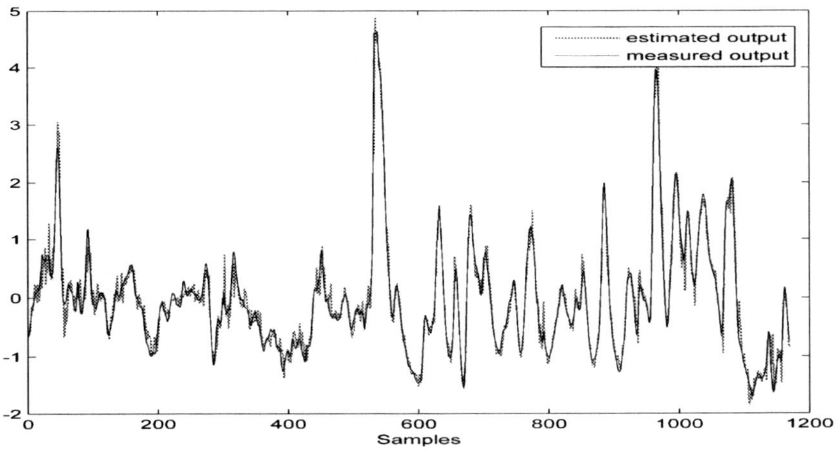

Figure 6.32. Comparison of measured and real-time estimation of C4 concentration in the bottom flow of the debutanizer column obtained using the enlarged one-step-ahead model

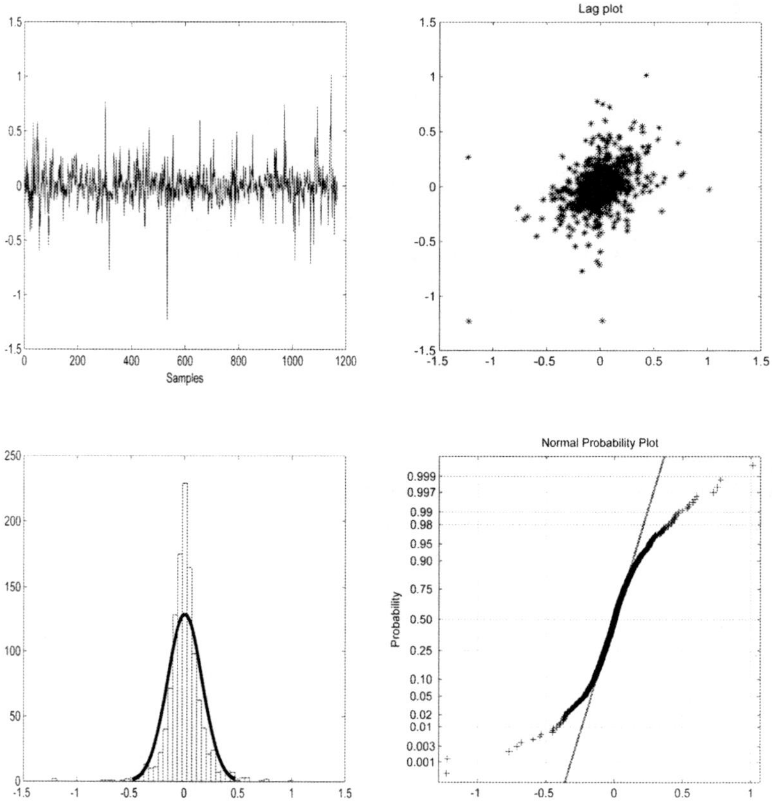

Figure 6.33. 4-plot analysis of the one-step-ahead enlarged model residuals

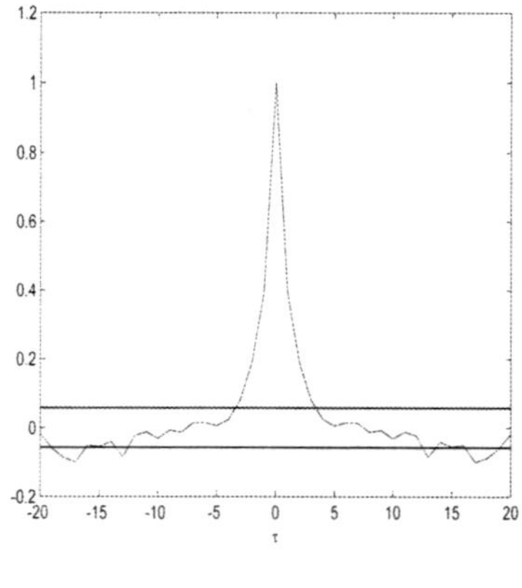

Figure 6.34. Autocorrelation function of Equation 6.2 for the model residuals

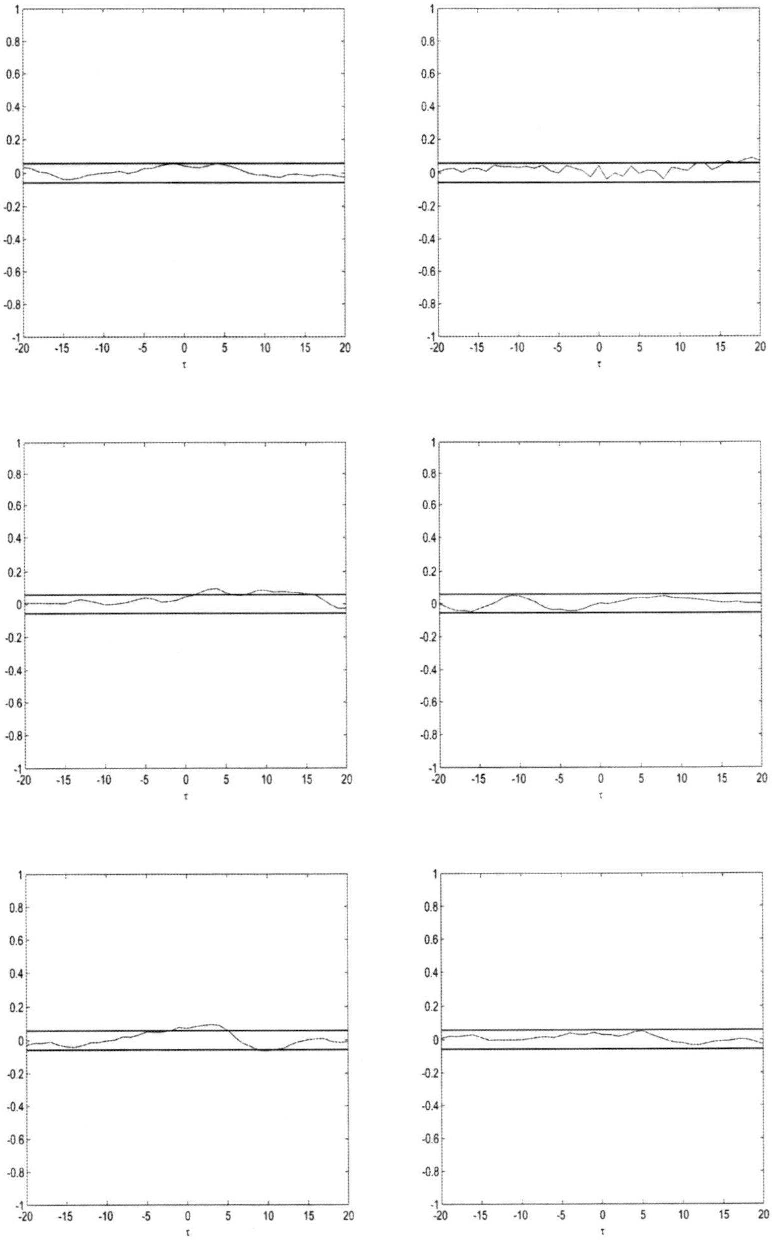

Figure 6.35. Correlation functions for Equation 6.3. Inputs from u_1 to $(u_6+u_7)/2$

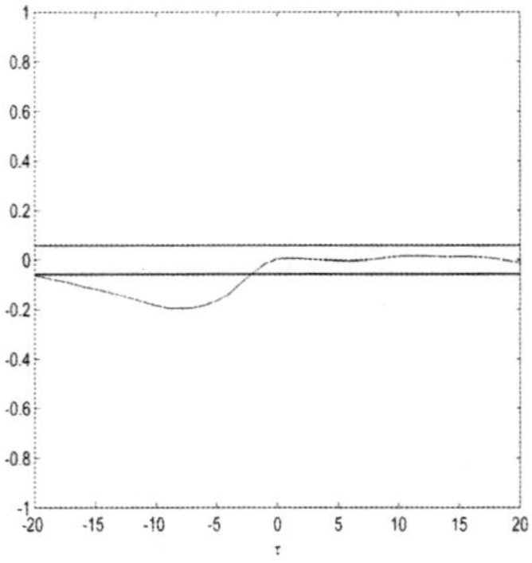

Figure 6.36. Correlation function for Equation 6.3. Plant output

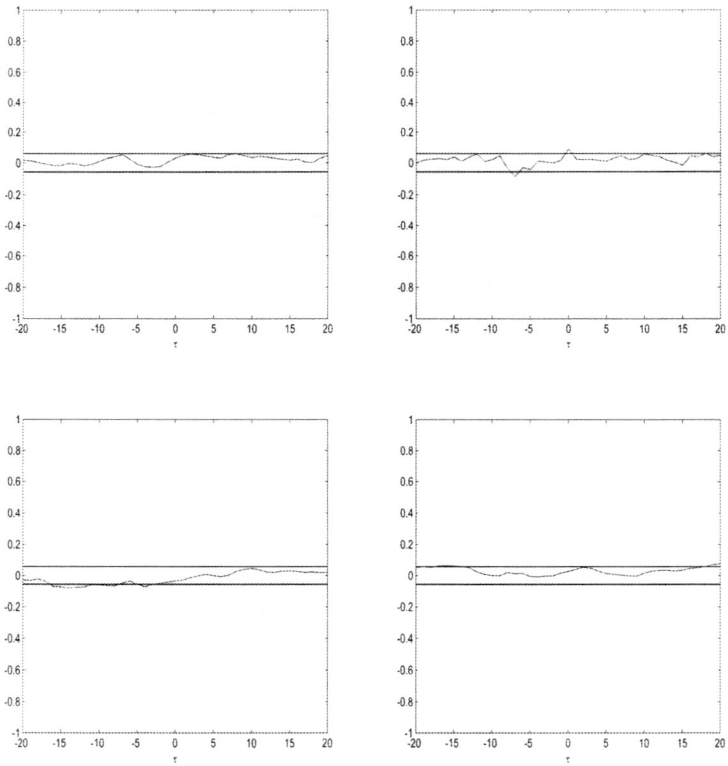

Figure 6.37. Correlation functions of Equation 6.4. Model inputs from u_1 to u_4

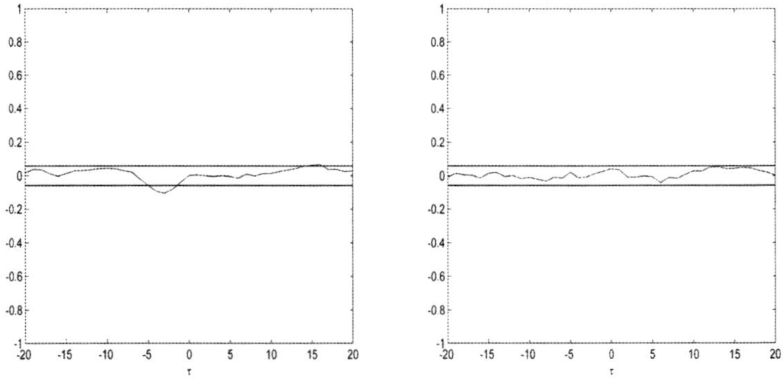

Figure 6.38. Correlation functions of Equation 6.4. Inputs u_5 and $(u_6+u_7)/2$

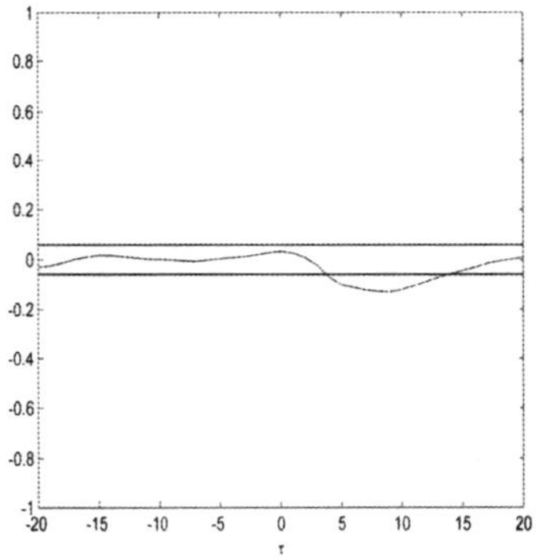

Figure 6.39. Correlation function of Equation 6.4. Plant output

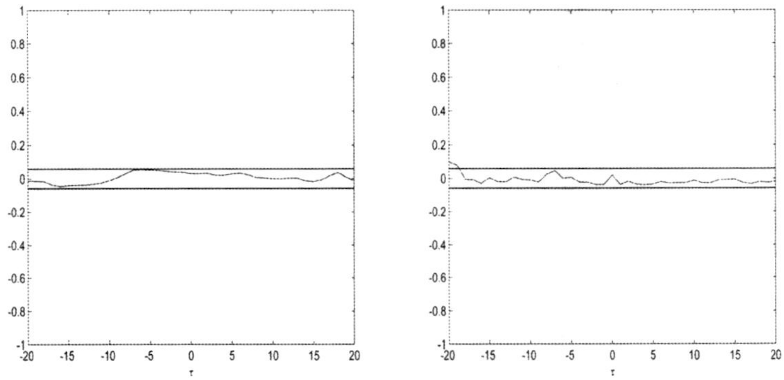

Figure 6.40. Correlation functions of Equation 6.5. Model inputs u_1 and u_2

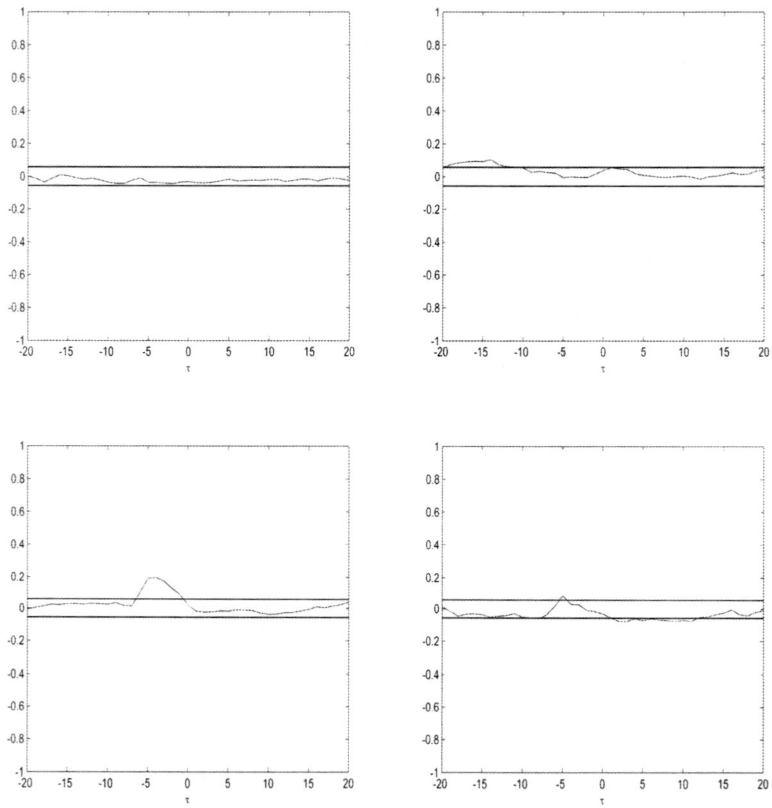

Figure 6.41. Correlation functions for Equation 6.5. Inputs from u_3 to $(u_6+u_7)/2$

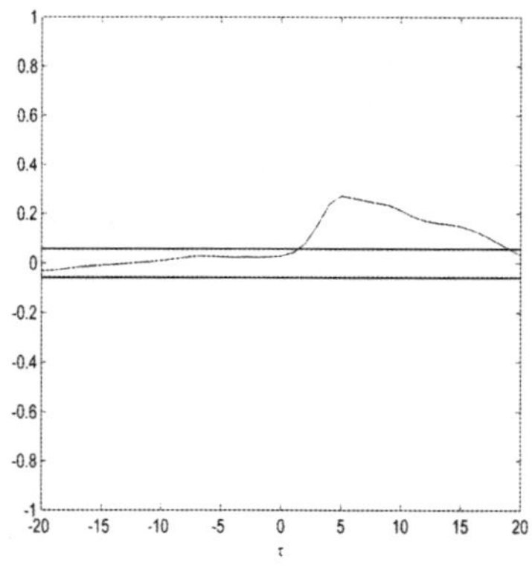

Figure 6.42. Correlation function of Equation 6.5. Plant output

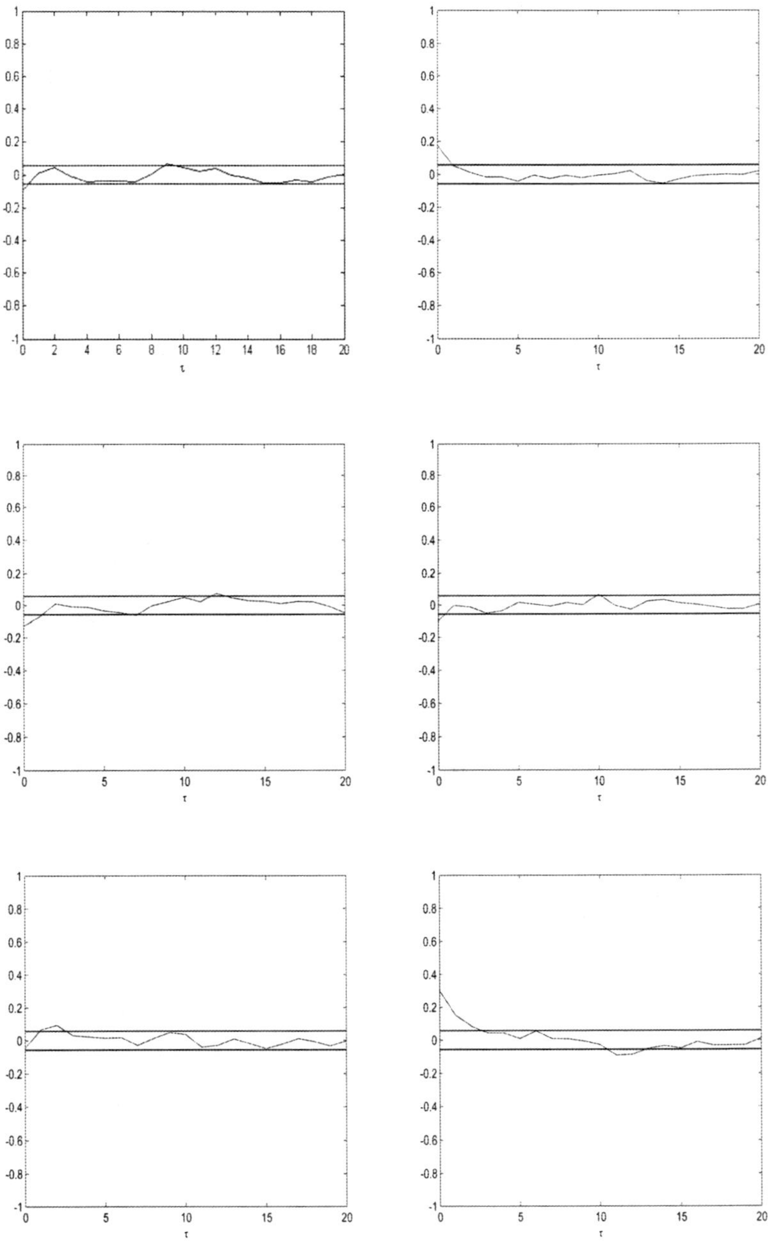

Figure 6.43. Correlation functions for Equation 6.6. Inputs from u_1 to $(u_6+u_7)/2$

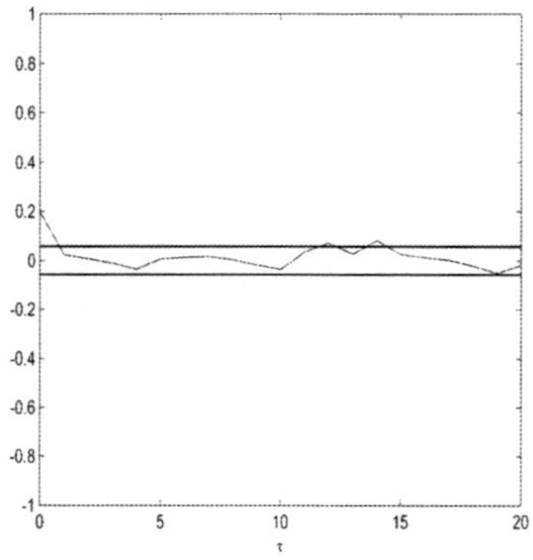

Figure 6.44. Correlation function of Equation 6.6. Plant output

Comparative analysis of the two one-step-ahead models allows some conclusions to be drawn.

As regards the residual trend, reported in the first sub-graph of the 4-plot analysis, the enlarged model shows a lower number of peaks, but suffers from some larger values (see also the corresponding lag plots and histograms).

In the correlation graphs for the enlarged model, the values outside the confidence interval are either reduced or have disappeared.

Though a slight improvement can therefore be argued in the soft sensor performance, it is not relevant, and the decision about the final model should be based on the on-line performance of the models and should, in general, also include considerations about model complexity.

6.5 Conclusions

Any data-driven model needs a careful validation phase performed observing the model behavior when processing “fresh” data. The very first and the most straightforward analysis is the graphical comparison of simulated data against acquired output variables. The choice among models that show similar behaviour can be obtained with further analysis of model residuals, as widely shown in this chapter. For linear systems residual autocorrelation and correlation analysis between residuals and model inputs can outline the presence of strong undesired correlations with lagged inputs that can be therefore included to improve model performance.

In the case of nonlinear models the designer should use a wider set of correlation functions, as indicated in this chapter.

The final step will be in any case the on-line validation of the soft sensor. This will require also a careful monitoring of input variable trends to avoid that large changes force the process to working points not considered during the design phase.

7

Strategies to Improve Soft Sensor Performance

7.1 Introduction

This chapter describes some strategies that can be used to improve the performance of soft sensors. Strategies typical of the soft computing approach to aggregate simpler computing structures are introduced for two different case studies. Both strategies use neural models as building blocks for the Soft Sensor design in two industrial applications, but they highlight different reasons for using an aggregation approach.

The first case refers to the soft sensor design for the SRU already considered in Chapter 5. The strategy introduced in Section 7.2 is intended to cope with the great number of sub-optimal neural models that are obtained, and generally rejected, during the design of a model. Using this approach, known as model stacking, sub-optimal models are aggregated with the hypothesis that the combination of a number of models can behave better that the fittest one used by itself.

In the second case, reported in Section 7.3, aggregation is the natural choice because the modeled system (the Powerformer Unit, already considered in Chapter 5) presents two different working points. Though a single nonlinear model could be designed to describe system behavior, a strategy based on the fusion of two different models can be adopted. In this case study, a fuzzy algorithm has been used to allow a smooth transition between the two models, in accordance with the system inputs.

The strategies described have been introduced with the aim of showing some of the possibilities offered by soft computing to improve soft sensor design. Of course, they do not cover all the possible techniques, but reflect personal experiences.

7.2 Stacked Neural Network Approach for a Sulfur Recovery Unit

A number of strategies are available to improve the performance of a soft sensor, based on combining a set of simpler models (Wolpert, 1992; Zhang *et al.,* 1997; Hashem, 1997; Zhang, 1999; Sridhar, Bartlett and Seagrave, 1999; Zhang, 2002; Ahamad and Zhang, 2002).

During the design of a soft sensor a number of trial models are usually obtained. Since each model can behave differently in different regions of the I/O space, a combination of some of these models can improve the overall prediction capability. A combination of different models is usually called "stacking". Stacked generalization is actually a generic term referring to any strategy for feeding information from one set of generalizers to another, before forming the final guess (Wolpert, 1992).

When, as in the present case study, a set of neural network models are combined, the combination is called a "stacked neural network". Stacked neural networks work by deducing the biases of the generalizers with respect to a learning set and generalizing in a second space whose inputs are the guesses of the original generalizers and whose output is the correct guess. The information fed to the net of generalizers comes from multiple partitioning of the learning set. Starting from this idea, several approaches have been proposed in the literature. In Zhang *et al.* (1997), data for building nonlinear models are re-sampled using bootstrap techniques to form a number of sets of training and test data. For each data set, a neural network model is developed. The models thus obtained are aggregated using principal component regression (PCR). The approach is used to design a soft sensor for a batch polymerization reactor. In Hashem (1997), an optimal linear combination of a number of trained networks is discussed along with two algorithms for selecting the component networks. The combination weights are selected to minimize the MSE with respect to the distribution of the model inputs. In Sridhar, Bartlett and Seagrave (1999) an information theoretic stacking (ITS) algorithm for combining neural models is proposed. The ITS algorithm does not require the form of the combination to be specified but uses the data itself to develop the appropriate combining rule. In Zhang (2002), a simple average of individual networks is used to accomplish a fault detection task. A comparison of several linear combining approaches is made in Ahmad and Zhang (2002) on nonlinear system modeling.

As a case study, in this section, modeling of the acid gases hydrogen sulfide (H_2S) and sulfur dioxide (SO_2) in the tail stream of the SRU described in the Appendix is considered. As described in Chapter 5, a sensor for the SRU has been designed using a number of different strategies. The performance of soft sensors implemented by using MLP and RBF neural networks, NF networks, and NLSQ have already been described, and it was concluded that MLP-based models gave the best results. For this reason, in this Section MLP-based models will used to design a stacked model by comparing a number of linear and nonlinear strategies.

Stacking of models can be performed using any kind of model that the designer considers of interest. In particular, models can differ in a number of aspects. Cases where models differ in structure (*e.g.*, linear, nonlinear, neural, fuzzy, etc.), input variables and/or regressors, data sets used for their identification, and so on, can be considered. Here MLPs with the same input structure as those considered in Chapter 5 and with two output neurons are considered. Neural networks differ in their hidden neuron number, initial weights and training sets.

Due to the requirements enforced on the tail gas composition, the models searched for were ones that are able to accurately predict peaks in SO_2 and H_2S concentration in the tail gas. This aspect will be taken into account in evaluating the performance of the different models.

To train the different neural models, a set of patterns acquired over a period of two months (sampling time $T_s = 1$ min), covering all the possible process working points, was considered. The network training used different sets of 1500 randomly selected I/O samples, while a different set of about 15000 data was used for testing and comparing the different models. All the networks were trained using the Levenberg–Marquardt algorithm, with the early stopping approach to prevent overfitting. A set of 200 different neural models was considered.

The networks were compared in terms of the correlation coefficients between actual data and corresponding estimations, on both learning and testing data, the residual trends and their distributions. The ten networks giving the best performance were considered for combination in the different stacking approaches. Table 7.1 gives the correlation coefficients between the actual and predicted outputs, computed over the set of 15000 testing patterns. Best values are reported in bold for each output variable. As can be observed from this table, the highest correlation coefficients are obtained with network 8 (with 8 hidden neurons), as regards H_2S prediction, and network 6 (with 7 hidden neurons), as regards SO_2 prediction.

Table 7.1. Correlation index of the ten best neural models for the SRU

	Corr_index H_2S	**Corr_index SO_2**
Net1	0.834	0.8027
Net2	0.8207	0.7989
Net3	0.8327	0.8050
Net4	0.7967	0.7873
Net5	0.8273	0.8038
Net6	0.8175	**0.8113**
Net7	0.8302	0.8000
Net8	**0.8397**	0.8059
Net9	0.7890	0.7932
Net10	0.8096	0.7808

Figure 7.1 compares the actual and predicted H_2S values for a subset of the testing patterns. The corresponding residual computed on the whole set of testing patterns is reported in Figure 7.2

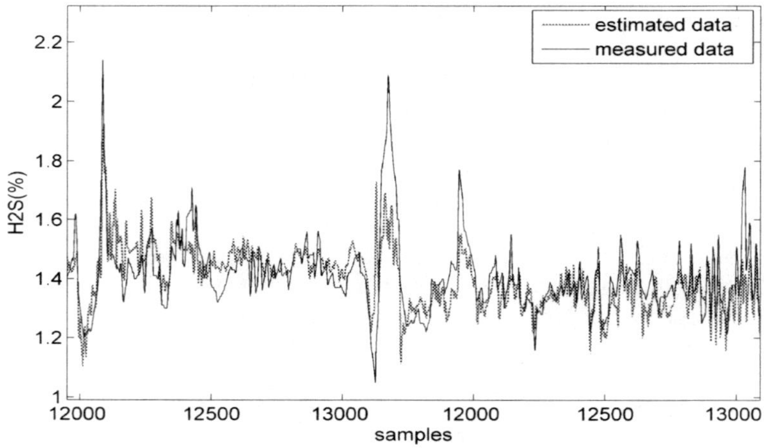

Figure 7.1. Performance of Net8 for the estimation of H_2S. Subset of validation data

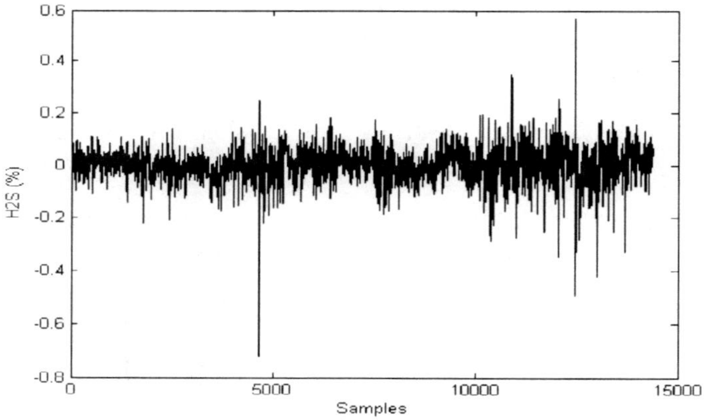

Figure 7.2. Residuals of Net8 for the estimation of H_2S

Figures 7.3 and 7.4 show the performance of Net6 for SO_2 prediction.

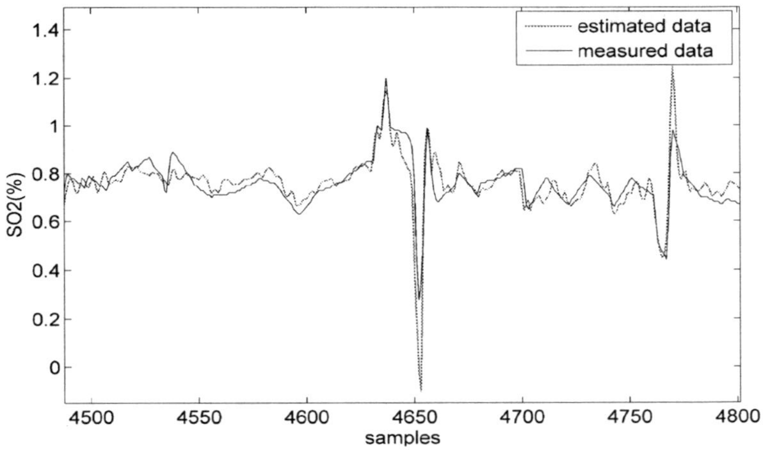

Figure 7.3. Performance of Net6 for the estimation of SO_2. Subset of validation data

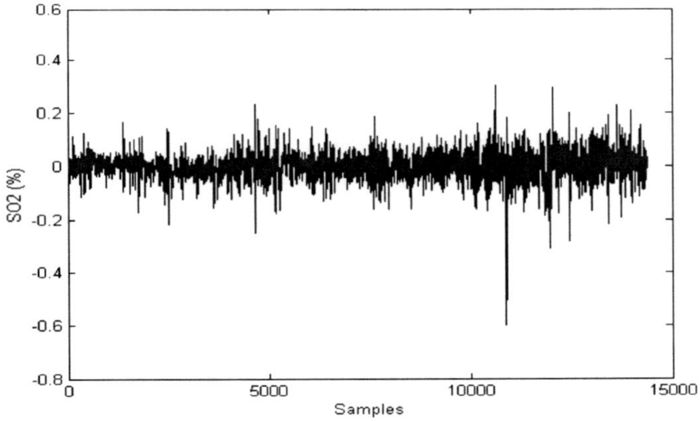

Figure 7.4. Residuals of Net6 for the estimation of SO_2

The overall performance of the networks is quite satisfactory, as shown by both the high value of the correlation coefficients and the graphic analysis. However, an in-depth comparison between the actual data and the corresponding estimations (Figures 7.1 and 7.3) shows, for both H_2S and SO_2, that even the best working networks are not able to accurately predict large peaks in test data. This drawback is also shown in Figures 7.2 and 7.4, where the residual vectors for the whole set of test data are reported.

In order to improve the prediction performance obtained, a set of stacking strategies were applied and a number of different stacking approaches were considered. Of course, the choice of predictors to be combined in the stacking strategy must obey some criterion, both as regards candidates and their number. Some criteria are reported in the work referenced at the beginning of this section.

A reasonable criterion is to select candidates on the basis of their performances as indicated by the correlation coefficient between the real system output data and the corresponding estimation. The number of candidates can be designed using a growing strategy in accordance with the correlation coefficients obtained, until no significant improvement is obtained.

In order to compare different stacking strategies, we decided to use a fixed number of models: the ten "best" neural predictors were used.

A scheme for a generic stacking structure is shown in Figure 7.5.

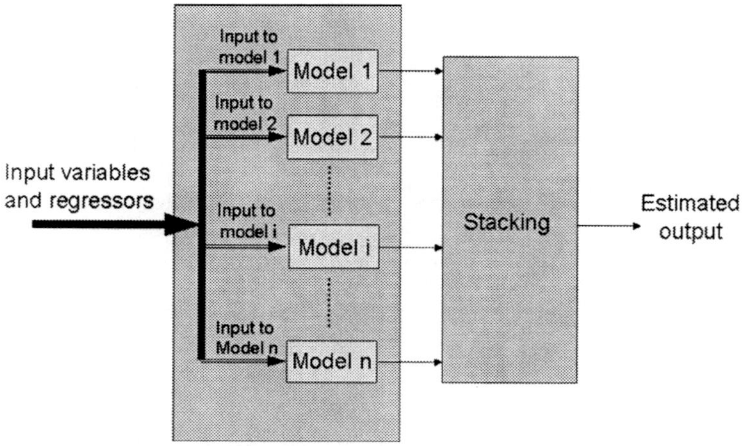

Figure 7.5. The stacking structure

Let $\hat{y}_i$ be the vector of the estimated output from the i_{th} neural predictor computed on a subset of patterns selected to compute the stacking parameters (in our approach a set of 1500 patterns was randomly selected from the whole data base). The corresponding estimation of the stacked network is, therefore

$$\hat{y}_{stack} = f(\hat{y}_1, \hat{y}_2, ..., \hat{y}_i, ..., \hat{y}_{10}, W) \tag{7.1}$$

where W is a matrix of weighting parameters to be searched for and $f(\cdot)$ is the stacking function. In the case of linear stacking approaches, Equation 7.1 reduces to

$$\hat{y}_{stack} = \sum_{i=1}^{10} w_i \hat{y}_i \tag{7.2}$$

The simplest approach suggested in the literature is based on a simple average of the available neural predictions. In this case, the weights w_i in Equation 7.2 are $w_i = 1$ ($i = 1,..,10$). The results obtained on the whole set of testing samples are given in Figures 7.6, 7.7, 7.8 and 7.9.

Figure 7.6 shows the stacked network prediction obtained, compared with actual data, on a subset of test samples for the H_2S concentration. The corresponding model residual is given in Figure 7.7. Corresponding Plots for SO_2 prediction are given in Figures 7.8 and 7.9.

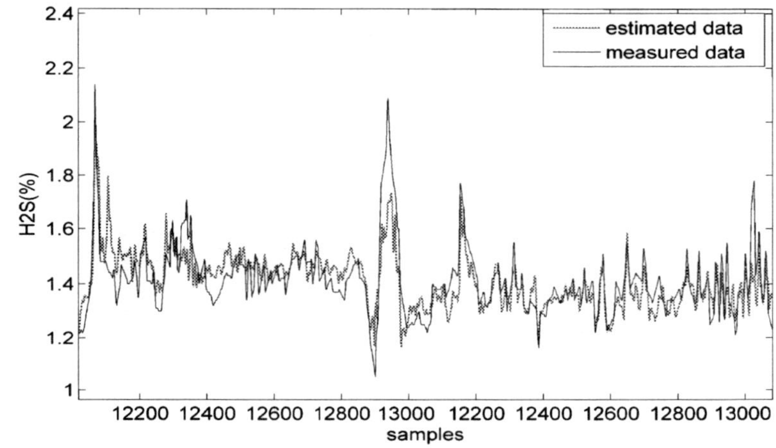

Figure 7.6. Performance of average stacking approach for the estimation of H_2S. Subset of validation data

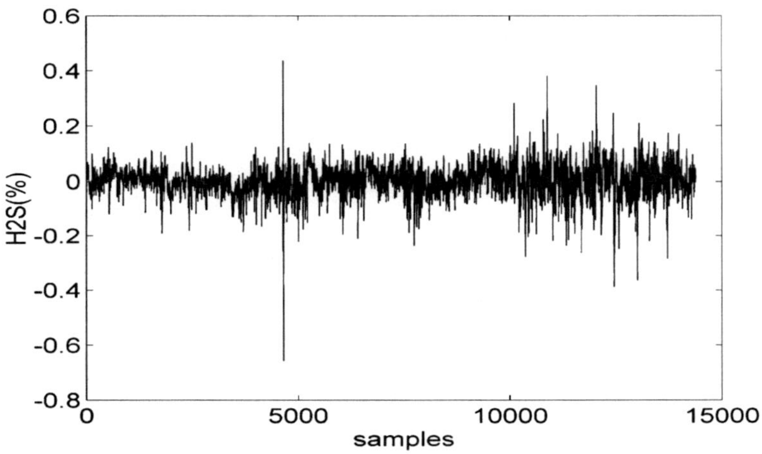

Figure 7.7. Residuals of average stacking approach for the estimation of H_2S

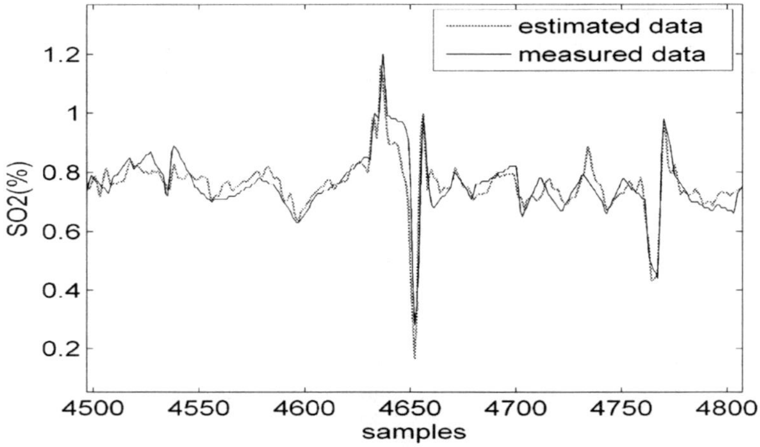

Figure 7.8. Performance of average stacking approach for the estimation of SO_2. Subset of validation data

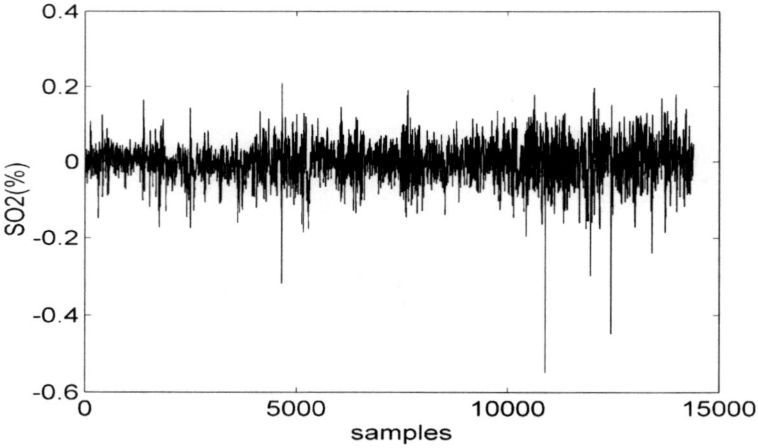

Figure 7.9. Residuals of average stacking approach for the estimation of SO_2

The MSE_OLC (mean square error_optimal linear combination) approach is based on deriving the weights w_i which minimize the MSE between the actual plant output and its estimation obtained by Equation 7.2, using a LMS approach. Let the observation matrix be

$$\widehat{Y} = [\widehat{y}_1 \ \ \widehat{y}_2 \ \ \ldots \ \ \widehat{y}_i \ \ \ldots \ \ \widehat{y}_{10}] \tag{7.3}$$

and

$$w = [w_1 \ w_2 \ \dots \ w_i \ \dots \ w_{10}]^T \tag{7.4}$$

the combining weight vector. It is computed as

$$w = \left(\hat{Y}^T \hat{Y}\right)^{-1} \hat{Y}^T y \tag{7.5}$$

where y is the vector of the actual plant output. As stated above, a set of 1500 samples was used to compute the weight vector. The results obtained on the whole set of 15000 patterns are given in Figures 7.10, 7.11, 7.12 and 7.13. These figures have the same meaning as in the case described above for the averaging stacking approach.

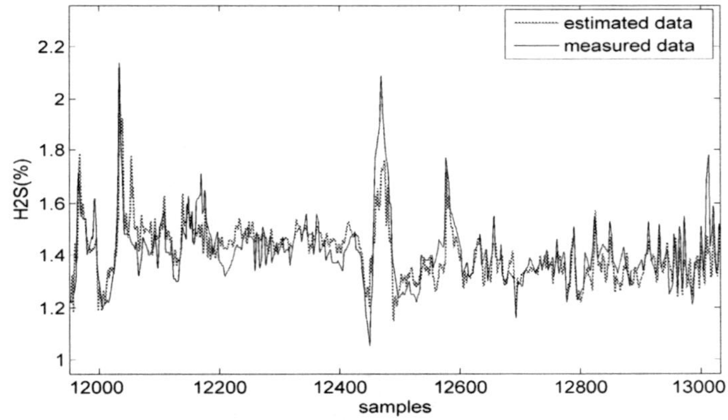

Figure 7.10. Performance of the MSE_OLC approach for the estimation of H_2S. Subset of validation data

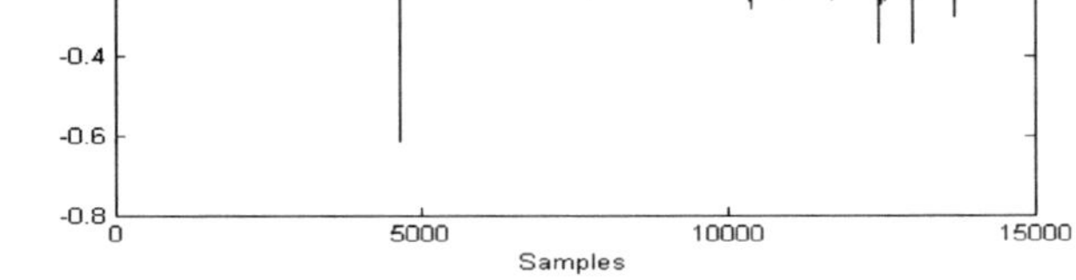

Figure 7.11. Residuals of the MSE_OLC approach for the estimation of H_2S

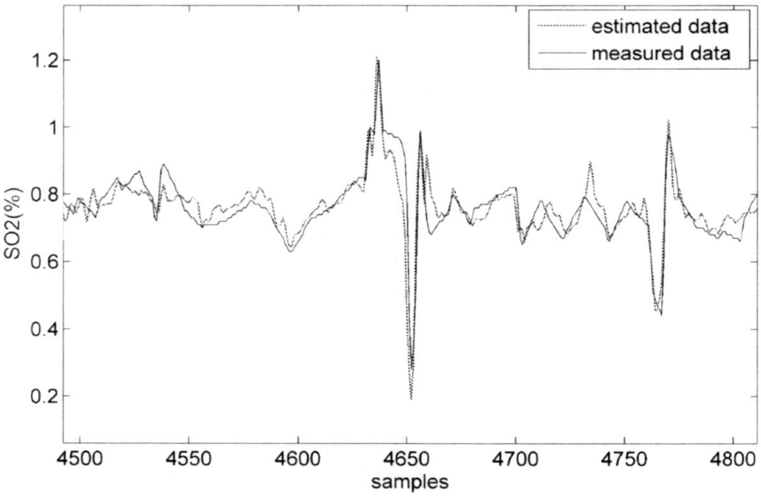

Figure 7.12. Performance of the MSE_OLC approach for the estimation of SO_2. Subset of validation data

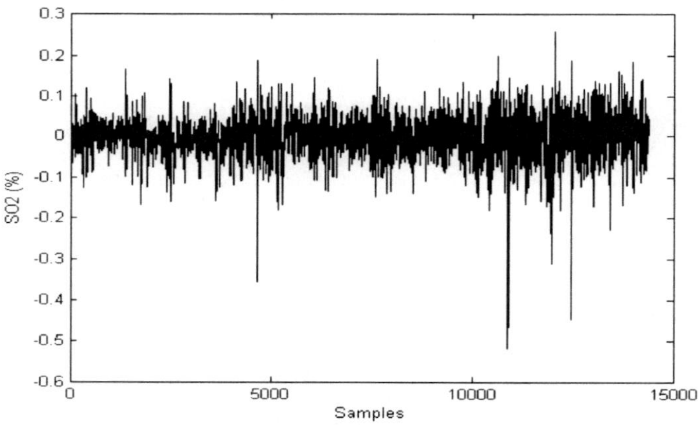

Figure 7.13. Residuals of the MSE_OLC approach for the estimation of SO_2

A further method, generally proposed in order to reduce the problem of data collinearity, is PCR. In this case, PCR is used as stacking strategy for the processing of SRU data. This approach is used to cope with possible drawbacks deriving from the highly correlated nature of the individual network predictors. Following this approach, the matrix $\hat{Y}$ is decomposed as a sum of a series of rank one matrices through decomposition of its principal components:

$$\hat{Y} = t_1 p_1^T + ... + t_{10} p_{10}^T \tag{7.6}$$

where t_i and p_i are the i_{th} score vector and the corresponding orthogonal loading vector. PCR allows the plant output to be estimated as a linear combination of the first k principal components of $\hat{Y}$. Let the loading matrix be

$$P_k=[p_1....p_k] \qquad k \le 10 \tag{7.7}$$

and the corresponding score matrix be

$$T_k = \hat{Y}P_k \tag{7.8}$$

The stacked estimation of the plant output is computed as

$$\hat{y}_{stack} = \hat{Y}w \tag{7.9}$$

and the combining vector w is obtained by using the LMS approach as

$$w = P_k\left(T_k^T T_k\right)^{-1} T_k^T y \tag{7.10}$$

In the application, the value of k was selected by considering the relative importance of each component with respect to the principal component. Four components were selected for H_2S and for SO_2. The results are given in Figures 7.14, 7.15, 7.16 and 7.18.

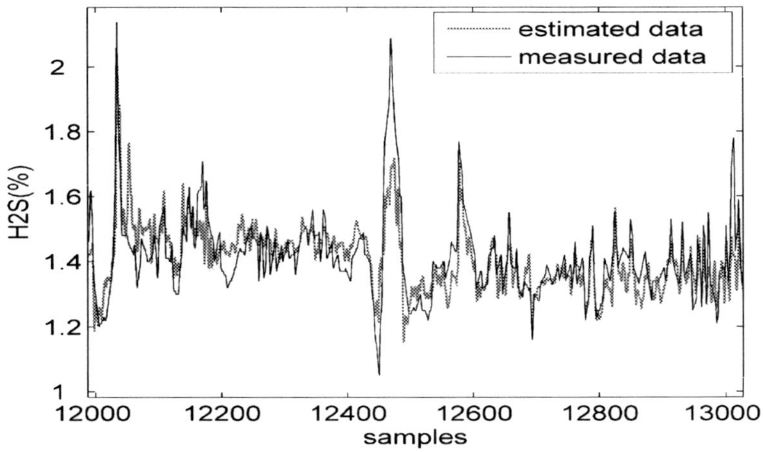

Figure 7.14. Performance of the PRC_OLC approach for the estimation of H_2S. Subset of validation data

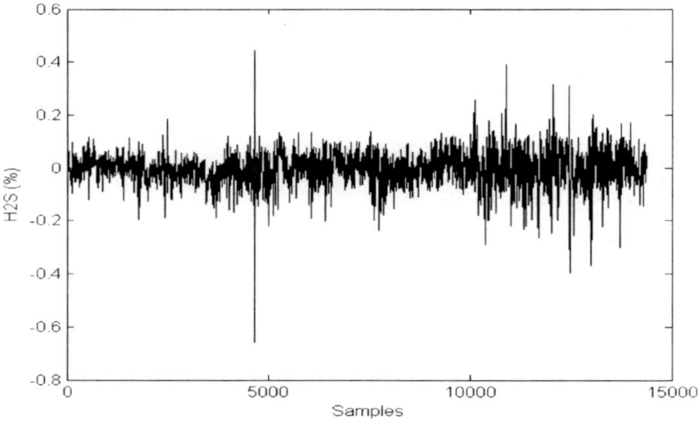

Figure 7.15. Residuals of the PCR_OLC approach for the estimation of H_2S

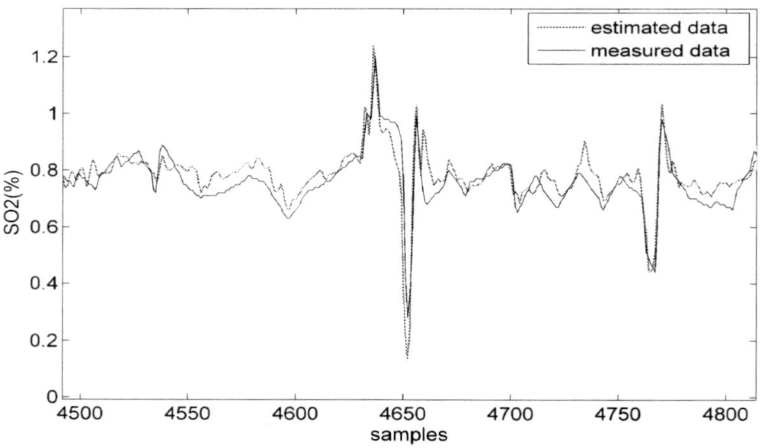

Figure 7.16. Performance of the PCR_OLC approach for the estimation of SO_2. Subset of validation data

A nonlinear stacking strategy which will henceforward be called the neural network optimal nonlinear combination approach (NN_ONLC) is proposed in what follows. It uses the interpolation capabilities of MLPs to combine the entries in $\hat{Y}$ to estimate the actual output y. The matrix $\hat{Y}$ is used to train a set of MLPs with a number of sigmoidal hidden neurons ranging in the interval [1..4], using the vector y as the desired output. Larger numbers of hidden neurons caused a degradation in the performance of the estimator. The performance obtained with the different second-level networks is shown in Table 7.2, in terms of the correlation coefficient between the actual and estimated plant output, computed on the whole set of patterns. Table 7.2 shows that the best results were obtained by using a network with three hidden neurons for the estimation of H_2S and two hidden neurons for SO_2.

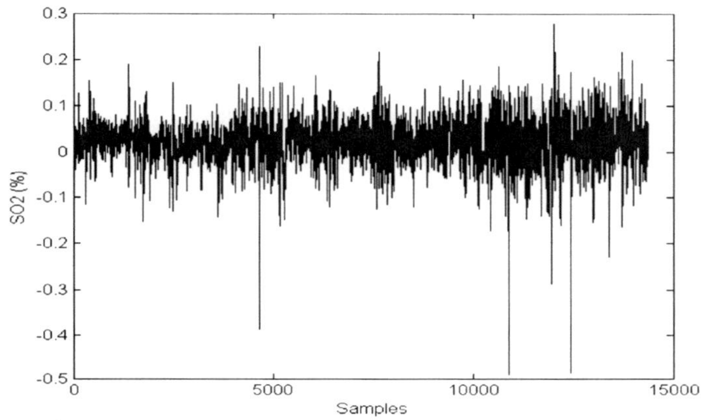

Figure 7.17. Residuals of the PCR_OLC approach for the estimation of SO_2

Table 7.2 Correlation index of the stacking approaches for the SRU

	Corr_index H_2S	**Corr_index SO_2**
Average	0.8619	0.8240
MSE_OLC	0.8638	0.8272
PCR_OLC	0.8620	0.8241
NN_ONLC_1	0.8632	0.8273
NN_ONLC_2	0.8637	**0.8301**
NN_ONLC_3	**0.8652**	0.8298
NN_ONLC_4	0.8642	0.8291

The results obtained using the best stacking networks are given in Figures 7.18, 7.19, 7.20 and 7.21.

A look through Tables 7.1 and 7.2 clearly shows, for the SRU application, the improvement obtained with stacking strategies compared with a simple neural network in terms of correlation coefficients; the various stacking strategies give comparable correlation coefficients. Slightly better results are obtained using NN_ONLC. MSE_OLC also gives good results, while PCR_OLC and simple average do not work so well. This general trend is confirmed for the estimation of both H_2S and SO_2.

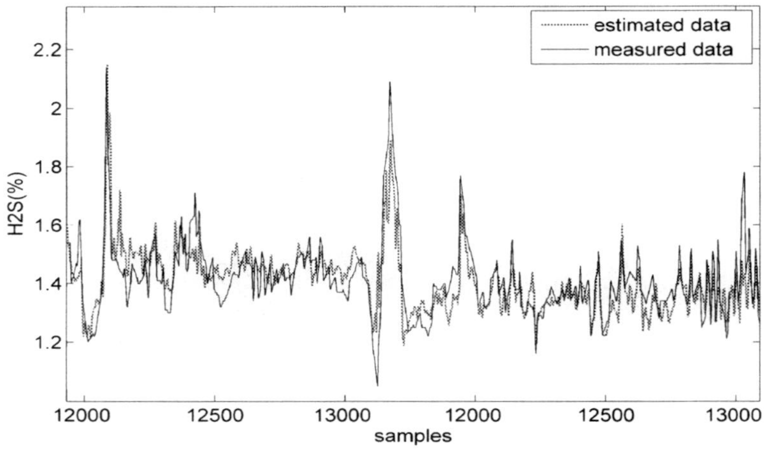

Figure 7.18. Performance of the NN_ONLC approach for the estimation of H_2S. Subset of validation data

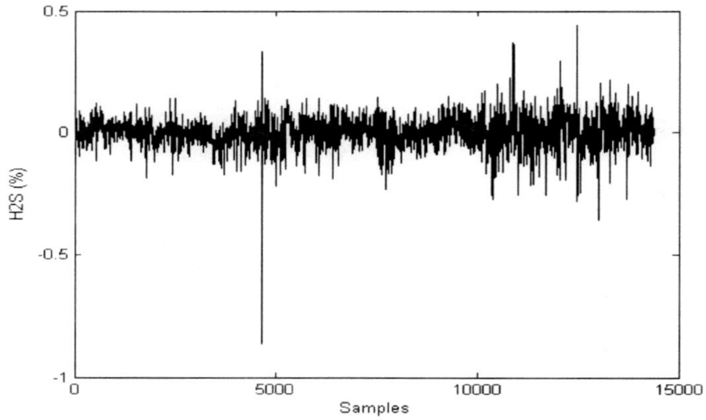

Figure 7.19. Residuals of the NN_ONLC approach for the estimation of H_2S

As mentioned above, stacking strategies were introduced to improve the ability of the model to estimate large data peak values. Hence, a careful visual inspection was carried out on the available data to evaluate this performance. Relevant trends are shown in the figures reported above and it is possible to observe that stacked networks work better than simple neural networks in the prediction of the two large peaks that occur in the interval between samples 12400 and 12600 for H_2S. Moreover, the proposed NN_ONLC is better able to predict the higher peak reported. This performance was consistently observed over the whole set of data, as can be observed from the residuals.

A similar behavior can be observed in the intervals between samples 4600 and 4700 and between 4750 and 4800 for SO_2.

From analysis of the results obtained, it can be observed that, though the different stacking strategies gave an improvement in the prediction capability, the differences between the various approaches are not sufficiently significant to establish the superiority of any one of them over the others. However, NN_ONLC was the strategy that guaranteed best results for both output variables.

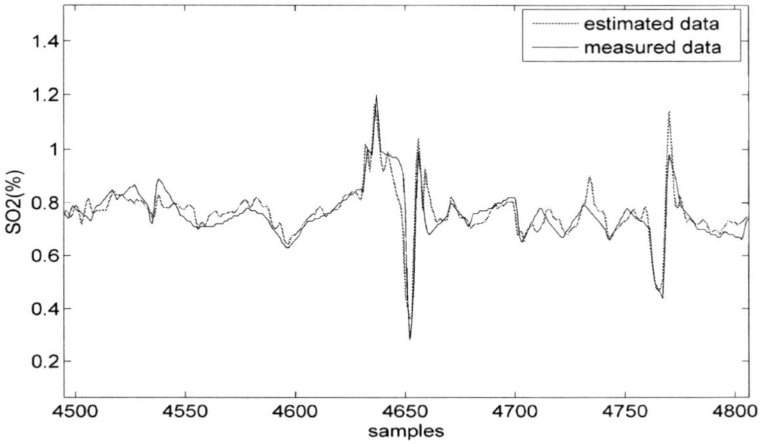

Figure 7.20. Performance of the NN_ONLC approach for the estimation of SO_2. Subset of validation data

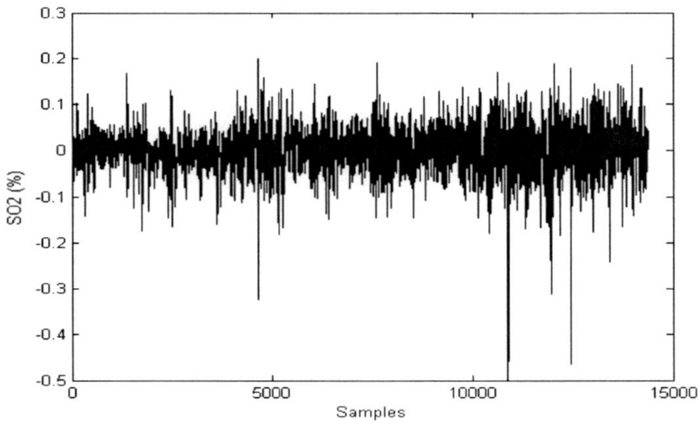

Figure 7.21. Residuals of the NN_ONLC approach for the estimation of SO_2

7.3 Model Aggregation Using Fuzzy Logic for the Estimation of RON in Powerformed Gasoline

Here the estimation of RON in powerformed gasoline is considered again with the aim of showing how two different models, derived for two different working points, can be fused together, to obtain a single soft sensor, capable of giving satisfactory performance over the whole system dynamic range.

In Chapter 5 a linear model was described for one of the two working points and it was outlined that further improvements were required, because of plant nonlinearity. Two possible strategies can be adopted: it is in fact possible either to design a nonlinear model covering the system dynamics or to aggregate to simpler nonlinear models, each one devoted to one working point, by a suitable algorithm. In particular, fuzzy logic is a natural choice for such an application.

Both models are based on the same input structure introduced in Chapter 5, replacing the linear combination of input regressors with MLPs.

While in the first approach a single MLP is trained with data covering both working points, in the second approach two different MLPs are trained to cope with each working point; they are then coupled with a fuzzy selection algorithm which allows a smooth transition between the different working conditions.

The performance obtained in the first case is reported in Figures 7.22 and 7.23.

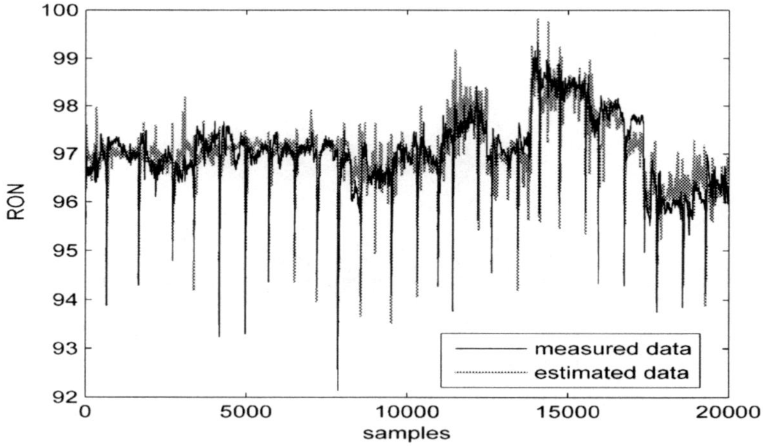

Figure 7.22. Performance of the MLP with 13 hidden neurons on a set of validation data covering both working points

The results reported were obtained with one MLP with 13 hidden neurons and represent the best performances obtained when trying this approach and MLP with different topologies. A correlation coefficient of 0.86 between the acquired RON data and estimated data was obtained.

As can be observed, model performance is improved with respect to linear models; moreover, a single model covers both working points, overcoming the possible difficulty of selecting a crisp threshold to assign data to each working point.

The second strategy is based on the fuzzy aggregation of two separate neural models. A typical trend of the input flow rate values, showing the two working points, is reported in Figure 7.24.

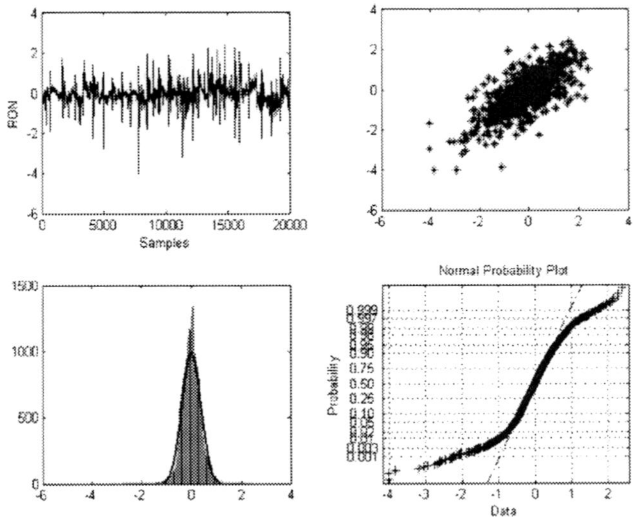

Figure 7.23. 4-Plot analysis of the residuals of the MLP with 13 hidden neurons on a set of validation data covering both working points

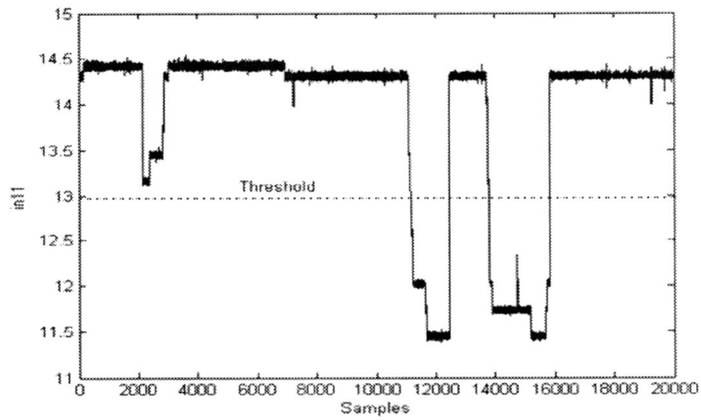

Figure 7.24. Time plot of input 11 used to discriminate the working points of the plant. Scaled data

The approach proposed requires, as a first step, that two separate neural models be trained to cope with the different working points. With this in view, the same pattern used to obtain the linear models, described in Chapter 5, was considered. For each neural model, a number of MLPs were trained in order to search for the number of hidden units.

In Figures 7.25 and 7.26, the performance of the network with 11 hidden neurons when processing a set of checking data for the first working point is shown. In Figure 7.25 the acquired RON values are compared with the nonlinear model estimations, while in Figure 7.26 the 4-plot analysis of the model residuals is reported. The correlation coefficient between measured plant values and estimated ones was 0.84.

In Figures 7.27 and 7.28 the performance of the network with 14 hidden neurons when processing the set of checking data of the second working point is shown. In this case, the correlation coefficient between measured plant values and estimated ones was 0.94.

The two neural models were then fused using a fuzzy algorithm designed with the adaptive fuzzy rule generator of MATLAB®. The algorithm uses as input the value of the variable named *in11*, while two membership functions (*in_11_low* and *in_11_high*) were designed in order to minimize the model residuals. The fuzzy algorithm has the following form.

- **RULE1:** `if` *in11* `is` *in_11_low* `then` *RON*=y_{model_1}
- **RULE2:** `if` *in11* `is` *in_11_ high* `then` *RON*=y_{model_2}

where y_{model_1} is the output of the first neural model and y_{model_2} is the output of the second neural model.

The defuzzyfied output is computed as

$$y_{fuzzy} = \frac{\mu_1\, y_{\text{model\_1}} + \mu_2\, y_{\text{model\_2}}}{\mu_1 + \mu_2} \tag{7.11}$$

where μ_1 and μ_2 are the activation levels of the two fuzzy rules.

In Figure 7.29, estimations obtained with the fuzzy-based soft sensor are reported. The 4-plot analysis of the corresponding residuals is shown in Figure 7.30. This approach guaranteed a correlation coefficient of 0.90 between acquired RON values and their estimation.

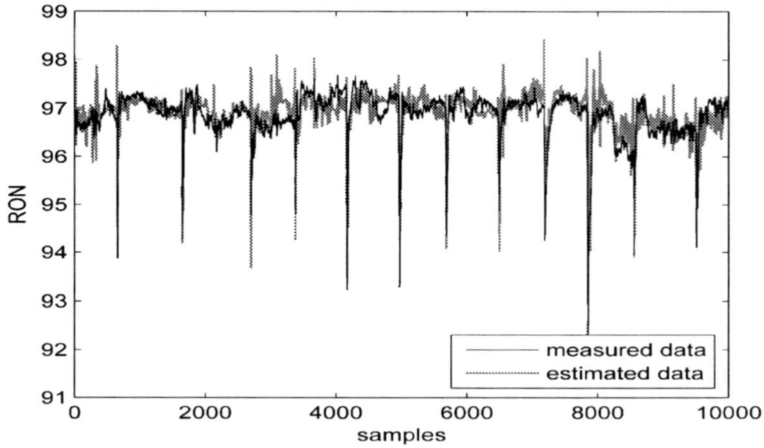

Figure 7.25. Performance of the MLP with 11 hidden neurons on a set of validation data covering the first working point

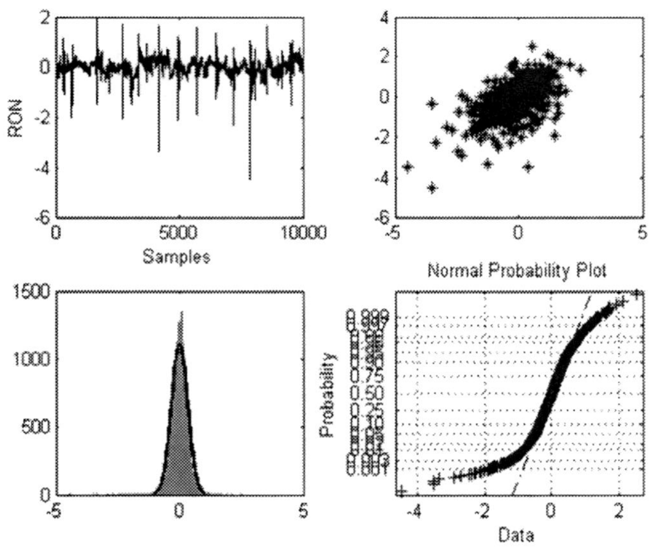

Figure 7.26. 4-Plot analysis of the residuals of the MLP with 11 hidden neurons on a set of validation data covering the first working point

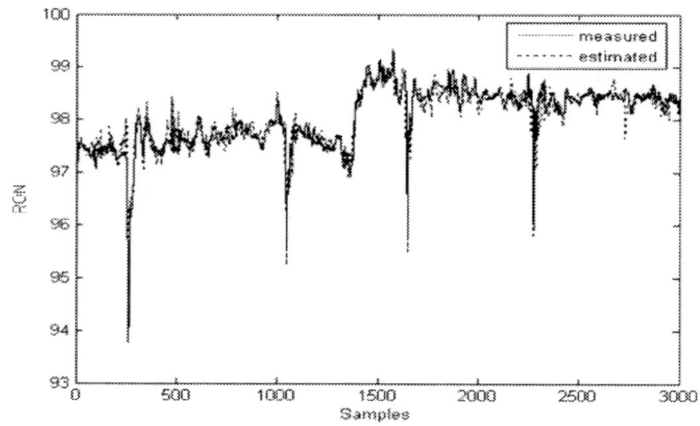

Figure 7.27. Performance of the MLP with 14 hidden neurons on a set of validation data covering the second working point

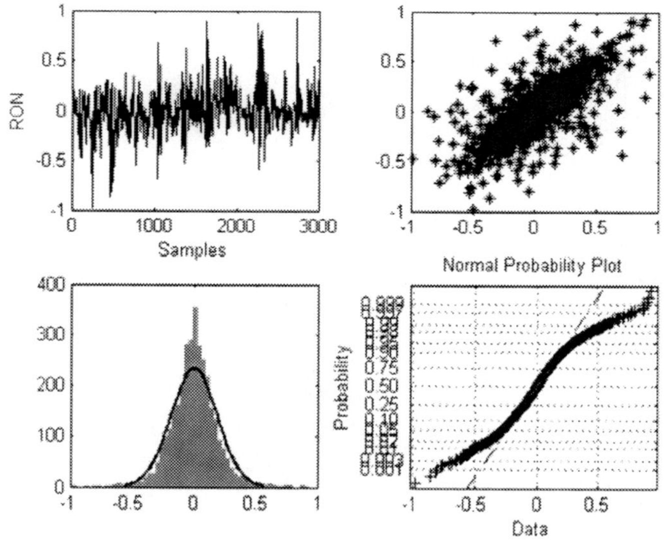

Figure 7.28. 4-plot analysis of the residuals of the MLP with 14 hidden neurons on a set of validation data covering the second working point

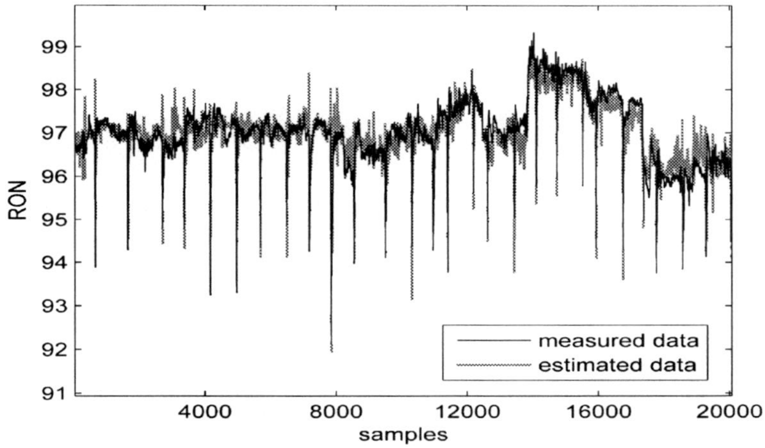

Figure 7.29. Performance of the fuzzy-based soft sensor on the whole validation data

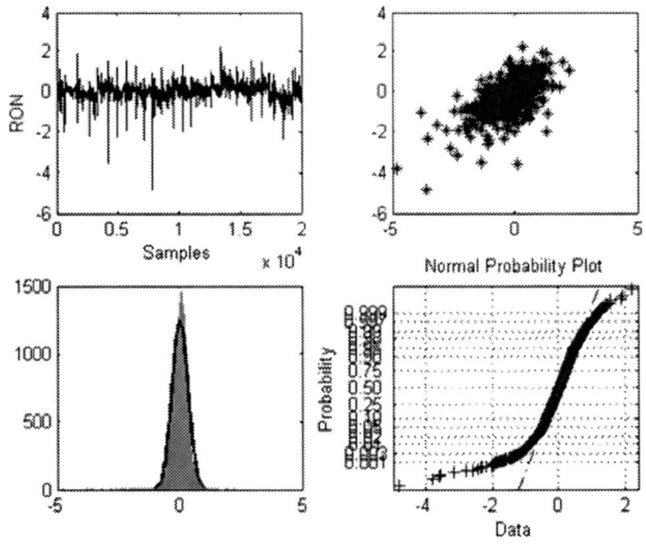

Figure 7.30. 4-plot analysis of the residuals of the fuzzy-based soft sensor on the whole validation data

The analysis of results reported in this section and in Chapter 5 confirms the superiority of nonlinear models with respect to linear LMS models for RON estimation. Also, among the nonlinear models, fuzzy switched models behave better than the single NN based model.

Further comparison of nonlinear models was based on their on-line performance. Taking into account a long period of on-line monitoring, it was decided by the plant technologist that the fuzzy approach gives better results.

As an example, in Figure 7.31 a comparison between actual data, the prediction obtained with the single neural model and the prediction obtained with the fuzzy switched model on data collected 3 months after on-line implementation is reported (only a subset of data is shown in the figure to emphasize the different model behavior).

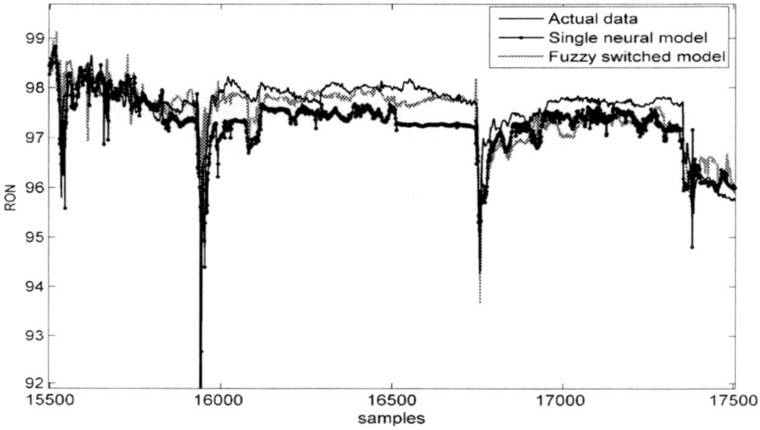

Figure 7.31. Comparison of soft sensors for RON estimation. On-line performance

The results reported in this chapter show that generally a suitable choice of aggregating strategy can improve soft sensor performance. Though the choice of the fusing strategy is strongly dependent on the application, and much responsibility is left to the designer, the possibility of obtaining better performance can be a good reason for spending more time on this matter.

7.4 Conclusion

Examples reported in this chapter have been selected to show the potentiality of soft computing to improve the performance of soft sensors. Soft sensor design is partially an artigianal activity which can be improved exploting *ad hoc* solutions for each problem. Neural networks, fuzzy logic, optimization strategies, classical identification strategies, and so forth should be considered as building blocks that can be used in the unifying framework of soft computing. This will adapt the used tools to the particular application and hence no general strategy can be suggested for this activity.

Stacking approaches, described in this chapter are still far from being a consolidated technology and a lot of theoretical open problems need to be investigated. As an example, given the set of designed models, no general consensus exists on the best method for the selection of candidates to be used in the stacked structure. Moreover, the aggregation strategy is an open issue and no unique methodology seems to exist that gives the best performance for all possible applications. The interested reader should use all available knowledge to adapt the strategies proposed in literature to solve its problem. References reported in this book can be used as a valuable starting point to this aim.

Notwithstanding the fuzziness of the topic, the experience of the authors, and results reported in literature show that the improvements can largely justify the time spent on this efforts. The cases of study reported in this chapter are just two significant examples showing the path to be followed and the improvements that can be obtained by using *ad hoc* stacking strategies.

8

Adapting Soft Sensors to Applications

8.1 Introduction

In the applications reported so far, a number of soft sensors have been described that were designed either for hardware back-up or for measuring device replacements during routine maintenance. Notwithstanding the possibility to use soft sensors for these reasons, other applications can be envisaged.

In Section 8.2, a VI is designed, to be used in the what-if analysis of the SRU considered earlier, *i.e.* to allow the controller designer to off-line simulate the system behavior with arbitrary inputs.

A different kind of monitoring is considered in Section 8.3 where a large urban area is considered in order to estimate the pollution level using soft sensors. In this case, the control policy is left to operators who, in accordance with local regulations, are required to perform some action. Also, the monitoring network has the form of a distributed system made up of a soft sensor network. In fact, as a result of the nature of the process, a model capable of estimating the pollution level at different points of the area considered was required, and this implied the use of space interpolation algorithms.

8.2 A Virtual Instrument for the What-if Analysis of a Sulfur Recovery Unit

In this section, a soft sensor is designed to perform what-if analysis on the SRU already considered for other kinds of applications.

In fact, so far the main objective of soft sensor design has been the back-up of available measuring systems. For this reason, NMA represented the most suitable choice. During maintenance of the gas chromatograph, output measurements are, of course, not available and ARX or NARX models are forced to work on the basis of infinite horizon prediction. This poses a high risk of poor estimation.

On the contrary, in the case of the what-if analysis, inputs are synthetic quantities, *i.e.* they are designed to analyze system reactions on a time span that makes sense, in accordance with system dynamics, so that finite horizon predictions are required. In this case, NARX models can be more efficient than the NMA ones used so far, because though any estimation error is fed back into the model, its magnitude is usually smaller. Moreover, finite-step-ahead prediction propagates model error effects only for a small number of iterations, which must, however, be carefully fixed by the designer.

The model used to perform the what-if analysis is therefore based on a cascaded structure made of one-step-ahead NARX models, each implemented in this application using an MLP. The cascaded structure is used to obtain a prediction up to 30 min ahead of phenomena relevant for establishing the working conditions of the process.

This VI was designed with the objective of simulating system dynamics for different control policies, to find efficient control strategies and/or to analyze effects of input variations (this can represent significant information, because some plant inputs come from other processes and are not totally under user control, nor predictable).

Details about the SRU plant are reported in the Appendix.

The first step to obtain the soft sensor for the what-if analysis is to build a one-step-ahead model.

Based on the consideration reported above, the regressor structure used for the NMA models might no longer be adequate for the NARX structure. In particular, models contain information about plant outputs at previous times. Moreover, since the outputs are correlated variables, both of them must be contained in the model input structure.

Due to the small number of candidate model inputs, a trial and error procedure was sufficient to determine a suitable model structure.

The sampling time used for the process was T_s=1.0 min. So that the requirement to predict system output up to 30 min would have required the use of a cascaded structure made up of thirty one-step-ahead predictors, and this was the first structure we considered. Though the one-step-ahead model performed very well, the behavior of the cascaded structure was totally unacceptable.

As a compromise, it was decided to reduce the total number of cascaded elements, by performing at each step the output prediction at step $k+2$. The following model structure was therefore used.

$$\begin{aligned}
[H_2S(k+2), SO_2(k+2)] = f\ [&in_1(k), in_2(k), in_3(k), in_4(k), in_5(k), \ldots \\
&\ldots, in_5(k-1), in_5(k-3), in_5(k-5), in_5(k-7), in_5(k-9), \ldots \\
&\ldots, H_2S(k-1), H_2S(k-4), H_2S(k-7), H_2S(k-10), \ldots \\
&SO_2(k-1), SO_2(k-4), SO_2(k-7), SO_2(k-10)]
\end{aligned} \tag{8.1}$$

This choice allowed the total number of MLPs in the cascaded structure to be reduced to 10. Each network has 18 input neurons, 4 hidden neurons, and 2 outputs. Figure 8.1 reports the performance of the model.

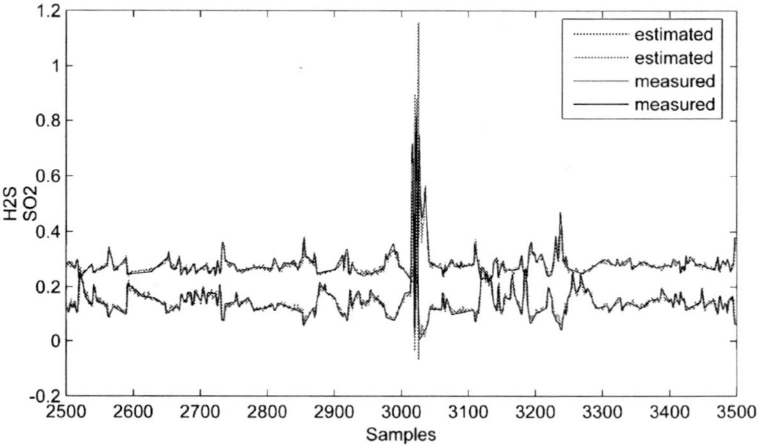

Figure 8.1. Performance of model reported in Equation 8.1. Subset of scaled test patterns

Model 8.1 was used to realize the cascaded structure that covers the required 30 min prediction interval. The scheme adopted is reported in Figure 8.2.

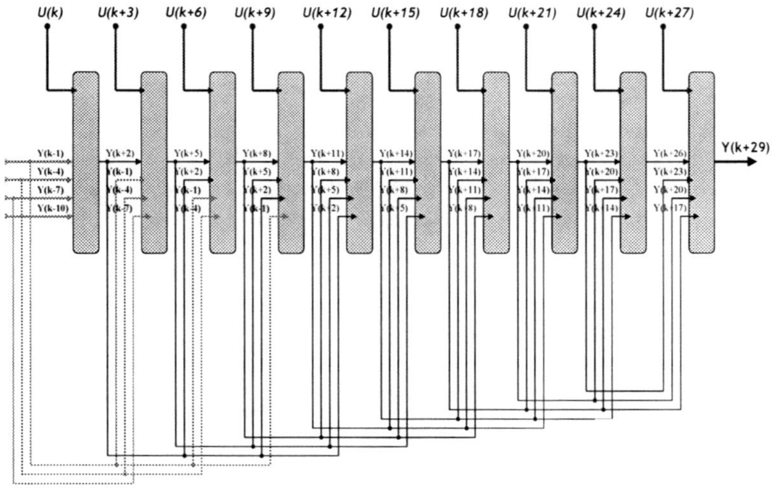

Figure 8.2. Scheme for the 30 min interval simulator. Output samples represented by a black line refer to estimated quantities. Gray lines refer either to historical data, available in the plant database, or to relevant initial conditions fixed by the user

From an analysis of Figure 8.2 it can be observed that the proposed structure allowed us to have the required system outputs at intermediate sampling times. It was therefore possible to evaluate the changes in the correlation coefficient between measured outputs and their estimation at each point in the structure.

Figure 8.3 reports the trends in the correlation coefficients for the two estimated outputs at each node in the cascaded structure. It can be observed that, generally, the larger the time delay, the smaller the corresponding correlation coefficient value. Anyway, it can also be observed that, up to thirty steps, the proposed model guarantees a satisfactory value of the correlation coefficient.

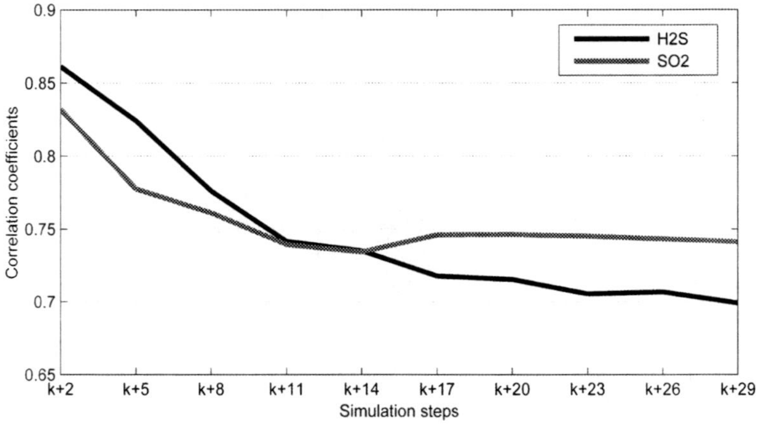

Figure 8.3. Trends in the correlation coefficients between measured output variables and their estimations *vs.* the simulation step

In order to emphasize the influence of the number of cascaded blocks, Table 8.1 reports the correlation coefficients between intermediate output estimations and measured values for both the structure with thirty cascaded blocks (each one performing a one-step-ahead prediction) and the structure reported in Figure 8.2.

Table 8.1. Comparison between the correlation coefficient at each step obtained using both 30 cascaded blocks and 10 cascaded blocks

	30 blocks		10 blocks	
	SO_2	H_2S	SO_2	H_2S
k+1	0.9785	0.9827		
k+2	0.8907	0.9157	0.8612	0.8321
k+3	0.7825	0.8361		
k+4	0.6916	0.7638		
k+5	0.6283	0.7139	0.8243	0.7775
k+6	0.5715	0.6704		
k+7	0.5175	0.6291		
k+8	0.4719	0.5901	0.7761	0.7612
k+9	0.4260	0.5460		

Table 8.1. (continued)

	30 blocks		10 blocks	
	SO_2	**H_2S**	**SO_2**	**H_2S**
k+10	0.3851	0.5040		
k+11	0.3447	0.4664	0.7416	0.7394
k+12	0.2844	0.4131		
k+13	0.2593	0.3836		
k+14	0.2982	0.3928	0.7351	0.7343
k+15	0.3116	0.3888		
k+16	0.3127	0.3755		
k+17	0.3065	0.3617	0.7177	0.7461
k+18	0.2863	0.3461		
k+19	0.2388	0.3219		
k+20	0.1821	0.2966	0.7154	0.7463
k+21	0.1613	0.2857		
k+22	0.1433	0.2739		
k+23	0.1311	0.2645	0.7054	0.7451
k+24	0.1322	0.2600		
k+25	0.1281	0.2516		
k+26	0.1223	0.2418	0.7067	0.7432
k+27	0.1191	0.2339		
k+28	0.1093	0.2263		
k+29	0.1004	0.2177	0.6990	0.7413
k+30	0.1009	0.2134		

It can be observed that even if the network performing one-step-ahead prediction guarantees a very high correlation coefficient, the error propagation through the 30 blocks causes rapid degradation of the estimation performances, and at step $k+5$ its performance is worse than those guaranteed by using simplified structures.

Once the 30-step-ahead predictor had been trained and its performance validated with respect to its capability of estimation up to 30 steps on the whole set of validation data, a further validation was performed, bearing in mind the objective of the soft sensor design.

As described above, the soft sensor was intended to help in the what-if analysis of the SRU and hence to analyze the system reaction to different input trends and/or control policies. In particular, a suitable control policy should avoid large peaks in both output variables. Bearing this in mind, available data were screened to search for relevant input trends that could be associated with the presence of large variations in the output variables. These subsets were used to perform simulations aimed at verifying the soft sensor capability to simulate output peaks. The simulations were performed using initial conditions both for inputs and outputs taken from historical data sets, not used in the learning phase, and comparing the simulated outputs with the recorded trends during the whole simulation period, lasting 30 min and corresponding to 30 steps, with a sampling time $T_s = 1.0$ min.

Figures 8.4 to 8.7 show examples of the simulation capability of the soft sensor. Figures 8.4 and 8.5 refer to the simulation of H_2S, while Figures 8.6 and 8.7 report examples regarding SO_2.

In these figures, the thin lines show the input trends recorded in the plant database, while the thick lines are the corresponding outputs. Solid lines are used for measured variables, while dashed lines refer to their estimations.

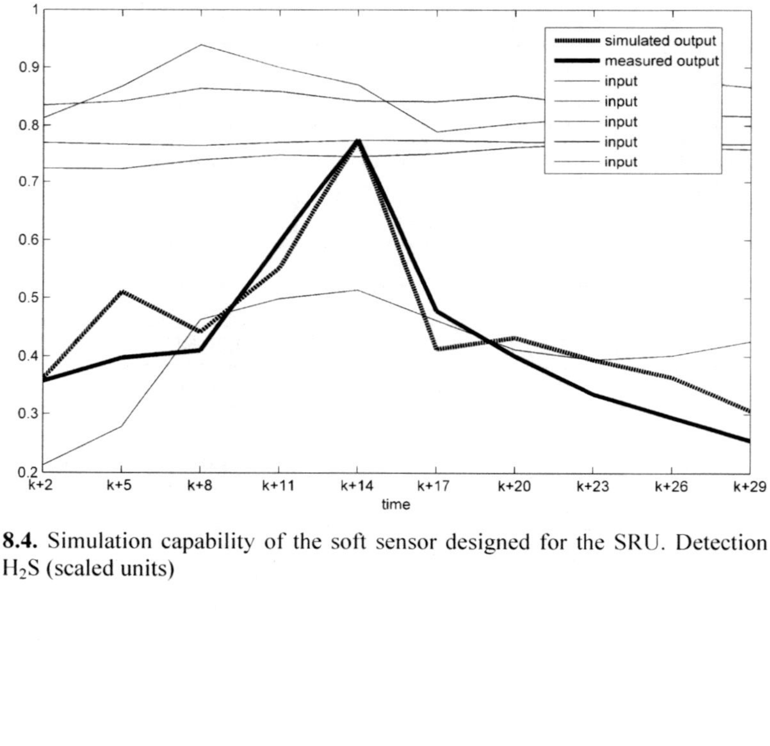

Figure 8.4. Simulation capability of the soft sensor designed for the SRU. Detection of a peak in H_2S (scaled units)

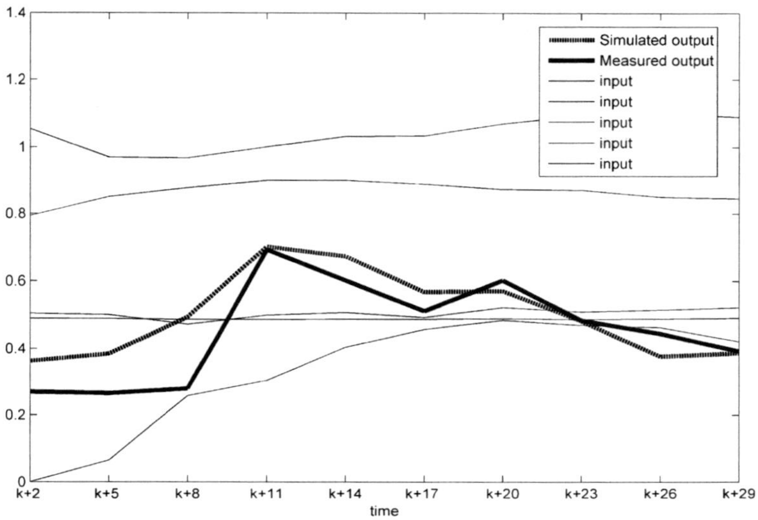

Figure 8.5. Simulation capability of the soft sensor designed for the SRU. Detection of a peak in H_2S (scaled units)

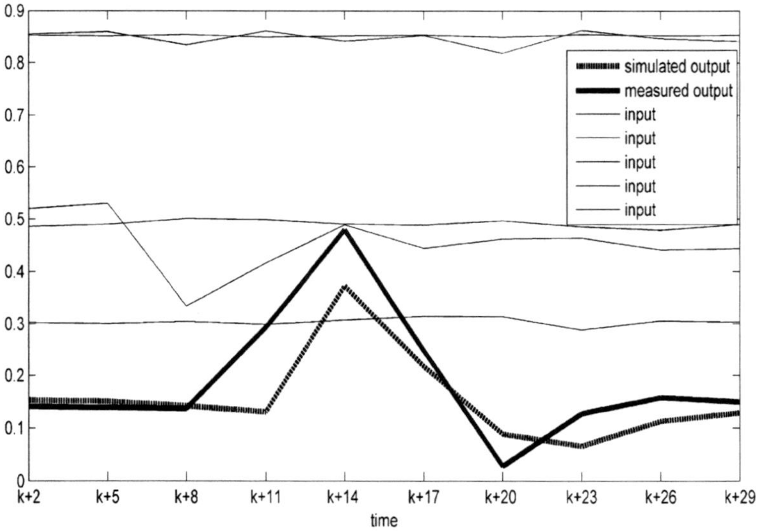

Figure 8.6. Simulation capability of the soft sensor designed for the SRU. Detection of a peak in SO_2 (scaled units)

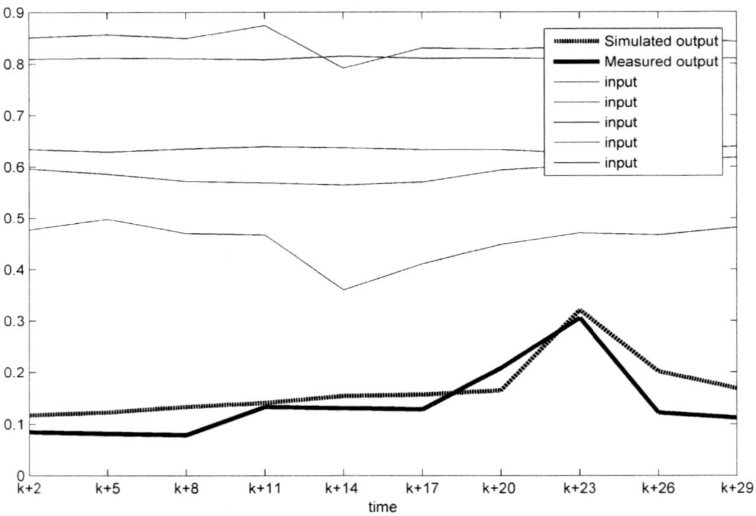

Figure 8.7. Simulation capability of the soft sensor designed for the SRU. Detection of a peak in SO_2 (scaled units)

The figures show that phenomena analyzed are extinguished in a time interval considerably shorter than 30 min, and this is evidence that the chosen simulation time span for the soft sensor suffices for the detection of peaks in the SRU output variables. Also the prediction accuracy in peak detection was considered satisfactory by plant technologists.

8.3 Estimation of Pollutants in a Large Geographical Area

Many human activities, both in urban and industrial areas, produce pollutants that can greatly modify the composition of the atmosphere. In urban areas, the main sources of air pollutants are motor vehicle emissions and domestic heating systems, while in industrial areas, production activities greatly contribute to emissions. Possibly, particular atmosphere conditions can produce thermal inversion phenomena that make it still harder for the air to mix.

In what follows, an application dealing with the problem of pollution level monitoring in urban areas using soft sensors is described. The interest in pollutant level estimation in large cities derives also from the development of regulations aimed at limiting each particular pollutant. Of course, different chemical substances will be considered in different ways, depending on the level of risk they represent for human health.

The very first attempts, in use for many years, tried by the public safety authorities were impositions of restrictions in emissions, *e.g.* by reduction in the number of vehicles allowed to circulate when a pollution level exceeding the maximum safety limit was detected (Stern, 1982).

Here the problem of the design of a monitoring network for air quality management is considered. The aim of the application described is the optimal exploitation of available measuring hardware, while a number of soft sensors use information redundancy to estimate pollutant levels at a set of sensitive points. The measuring system is, therefore, in the form of a distributed network, where each node is a monitoring station, and two different tasks need to be solved:

- to find the optimal observation points where it makes good sense to install available hardware;
- for each of these points, to design a soft sensor capable of predicting the pollution level as a function of relevant inputs.

Techniques for the optimal allocation of monitoring stations have been developed. A methodology for the optimization of the number of stations in a monitoring network based on a recursive procedure, is described below (Andò *et al.,* 1999).

A "first trial" monitoring network is assumed to be available, and soft sensors are designed to estimate the level of a pollutant at a monitoring point as a function of the pollutant levels recorded at other points. As a by-product, the sets of monitoring stations that supply redundant data and the smallest number of monitoring stations that must be left "alive" are determined. Stations that can be turned off, on the basis of previous analysis, become available for reallocation.

Now it is possible to start the second phase of the monitoring network design procedure: the most "significant" places for the reallocation of the monitoring stations in the area under consideration are therefore searched for.

In what follows, the application of the procedure described to a case study is reported.

As usual, the problem of outlier detection should be addressed. Here the most common problems generating outliers were detected during both the transmission and the auto-calibration phases of the measuring instrumentation.

Also in this case, expert knowledge was a precious source of information for the critical analysis of available data (*e.g.* experts had established that carbon monoxide levels in urban areas fall within the range 0.3 ppm to 40 ppm, and data outside this interval were cut out).

Further information about data significance can be obtained if the redundancy of available data, due both to the multi-input structure of the measurement systems (each station collects data on several pollutants) and to the distributed nature of the phenomena under investigation, is taken into account. The correlation among the levels of different pollutants measured in one station, and the correlation among the levels of one pollutant, recorded in different stations, can be hypothesized. Therefore, a critical analysis of the significance of data can be improved by using the cross-correlation technique.

The following procedure was suggested to look for anomalies at any station.

- For each station, and for each pollutant recorded at that station, compute the correlation coefficients between each pollutant and the others. Anomalous values for all correlation coefficients, related to one pollutant, allow the detection of any device that is not working properly in the station.

- Compute now the cross-correlation coefficients between the time series collected from the suspect hardware and the corresponding series acquired in other stations.

If both investigations provide low values of the correlation coefficients, the set of data regarding the pollutant is eliminated.

Examples of time plots of typical urban pollutant levels, after removing outliers, are reported in Figures 8.8 and 8.9. Five hundred samples per hour were acquired for both CO and SO_2. The corresponding mean hourly values are reported in Figure 8.8 for CO, while for SO_2 mean daily values are shown in Figure 8.9. Such a difference depends on the different way in which limits are set for these two pollutants by Italian law.

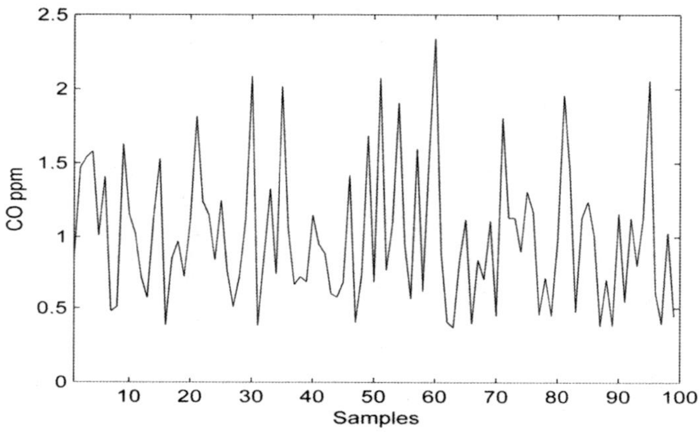

Figure 8.8. CO levels recorded by one measuring station in a typical medium size town, after outliers elimination

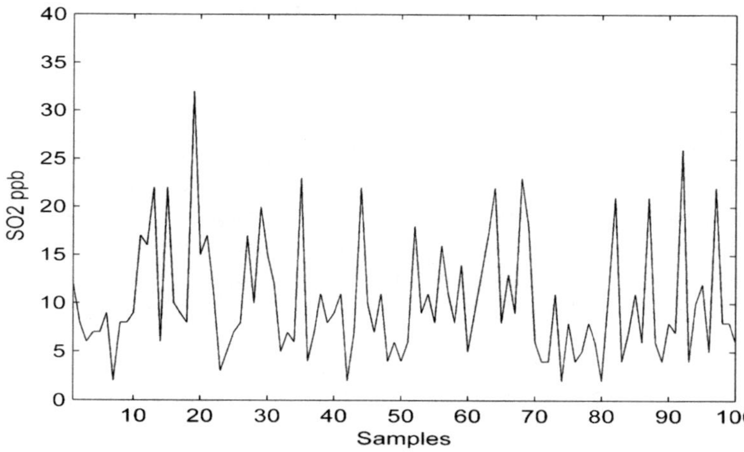

Figure 8.9. SO_2 levels recorded by one measuring station after outliers elimination

Once data are processed to detect and eliminate outliers, due to hardware malfunctioning, the optimization procedure for measuring station re-allocation can be started (an initial guess monitoring network structure is assumed to exist). The procedure is based on the use of mathematical models that allow the vector of pollutant values in one station to be estimated as a function of corresponding vectors recorded from other stations in the network (Andò *et al.*, 1999):

$$P_i^{ST_j} = f\left(P_i^{ST_1}, P_i^{ST_2}, \ldots, P_i^{ST_m}\right) \tag{8.2}$$

where $P_i^{ST_j}$ is the *i*th pollutant concentration value, recorded by the *j*th station considered and represents the output of the candidate soft sensor, m is the number of neighboring stations chosen between available monitoring stations and $f(\bullet)$ is an unknown function describing the model.

As regards the choice of the input stations m, a combinatory optimization problem must be solved, also considering that such stations are not necessarily those closest to the *j*th station. It was in fact observed that a more reliable tool for the selection of neighboring stations is represented by the correlation coefficient: cases were observed of measuring stations that guaranteed a larger correlation coefficient than closer stations, though they were farther away.

Once this step is accomplished, the structure of the $f(\bullet)$ function needs to be solved. In this application, nonlinear models based on neural networks were used. Nonlinear static models based on MLPs with one hidden layer were proposed. One MLP with one output neuron, corresponding to the pollutant level of interest, was considered for each pollutant recorded in each station.

To evaluate the performance of the models, both the mean value and the standard deviation of the model residuals were estimated.

Models 8.2 allow us to find stations that can be substituted by a soft sensor capable of estimating the pollutant considered .

If all soft sensors associated with pollutants recorded at one monitoring station work properly, that monitoring station is a candidate to be turned off, because measuring devices can be substituted by mathematical models. In order to be able to turn off such a station it is also necessary to check that data recorded by this station are not used as independent variables by other soft sensors.

When the procedure described ends, the set of available monitoring stations will be either removable, if all soft sensors designed for this stations perform adequately and data acquired by measurement instruments installed in the station are not used by any of Models 8.2, or else non-removable, if not replaceable with a mathematical model, because the performance index value is worse than the fixed threshold, or because data acquired by this station are already used by other soft sensors in the form of Equation 8.2.

Stations that have been declared removable become available for reallocation. This problem must be faced by considering the spatio-temporal nature of the problem.

With this in view, a family of continuous maps covering the urban area for the whole set of pollutants is estimated on the basis of data recorded at a finite number of points. Each map in the family represents a snapshot of the spatial distribution of a pollutant concentration in the urban area at a given time. This family of maps is analyzed to look for points where the available stations should be reallocated. Points where at least one pollutant concentration reaches the maximum value will be considered as candidates for station reallocation.

The continuous maps can be obtained from the recorded data using a suitable interpolation method. Here, the spline interpolation method (Kreyszig, 1999; Ahlberg, Nilson and Walsh, 1967) for two-dimensional data is applied.

A case study of the described procedure was introduced in Andò *et al.* (1999). There, data recorded using a monitoring network working in the city of Catania and intended to monitor the intra-urban air pollution due to a number of pollutants were considered.

Catania is a medium-sized town in southern Italy. It represents a typical European town, with a very congested downtown section and a large suburban area. Moreover a great number of people commute to the town each day in private cars. The problem of air pollution due to exhaust gases produced by motor vehicles is therefore very serious and the monitoring network was installed according to Italian law.

The monitoring network used to acquire data considered in the application described consisted of 22 stations for the acquisition of data on the following pollutants:

- carbon monoxide, CO;
- nitrogen dioxide, NO_2;
- sulfur dioxide, SO_2;
- hydrocarbons, NHMC.

Meteorological data were also acquired in order to take into account the possible influence of meteorological conditions on air pollution evolution.

The network has been operating since the beginning of 1993. Data collected from October 1993 to March 1994 were used for the case study, along with hourly acquired data. During this time, 15 stations out of 22 were working.

For some pollutants, Italian law refers to the average daily value, while for others the hourly value must be taken into account. For this reason, models based on hourly samples were considered for both CO and NO_2 while NMHC and SO_2 were modeled using daily values.

MLP neural models with four inputs were chosen as a compromise between the contrasting needs to obtain good soft sensor performances and as simple as possible mathematical models. Models were considered to be acceptable if they guaranteed an error index less than a threshold fixed for each pollutant.

The application of the described procedure allowed two stations out of the 15 working stations to be declared removable, and then substituted with the corresponding soft sensors. The other stations were declared non-removable, because either they featured performance index error values higher than the allowed value, at least for one pollutant, or data they acquired were used as inputs to soft sensors.

In Figures 8.10 to 8.13, performance obtained by using the designed soft sensors for one of the two removed monitoring stations are reported. In particular, the level of the four pollutants mentioned above, estimated using the neural-based soft sensors, are compared with the corresponding acquired data.

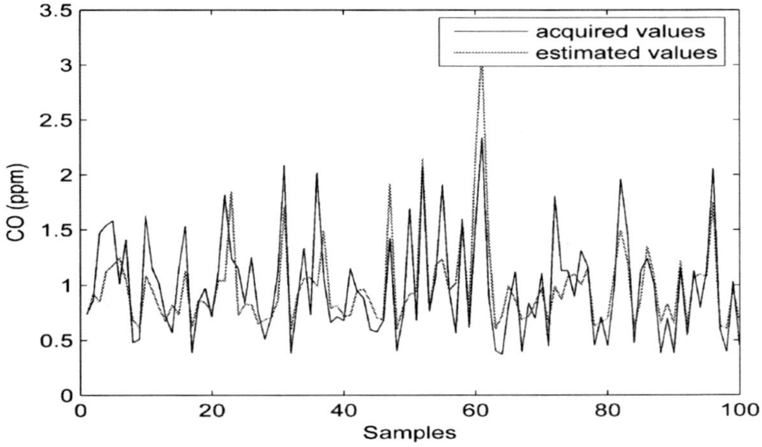

Figure 8.10. CO level (in ppm) estimated using the neural network based soft sensors for one of the removed stations

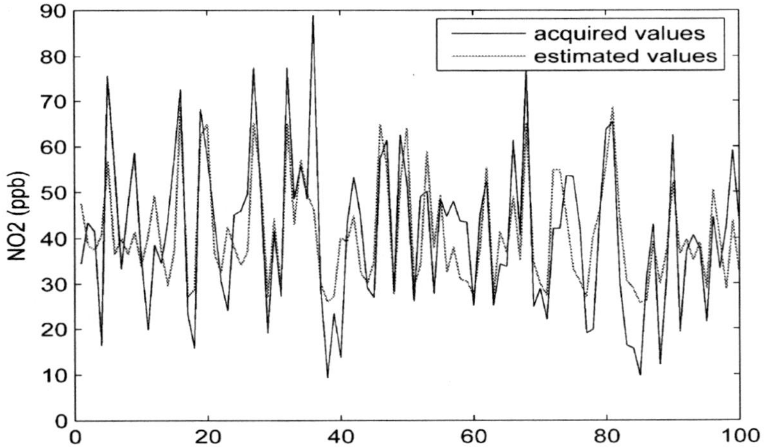

Figure 8.11. NO_2 level (in ppb) estimated using the neural network based soft sensors for the removed station

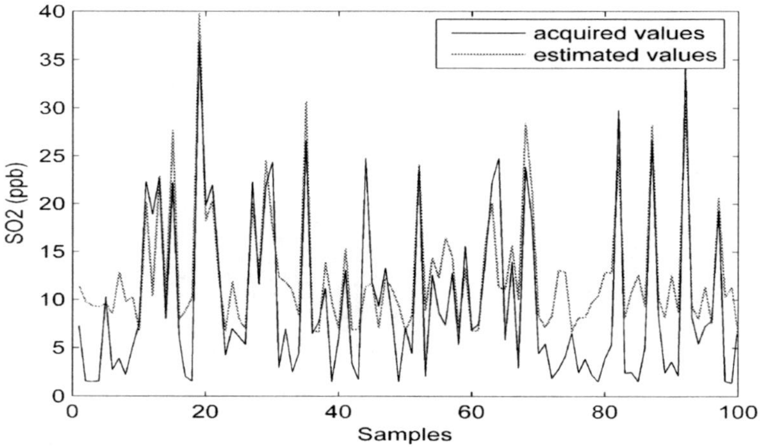

Figure 8.12. SO_2 (in ppb) level estimated using the neural network based soft sensors for the removed station

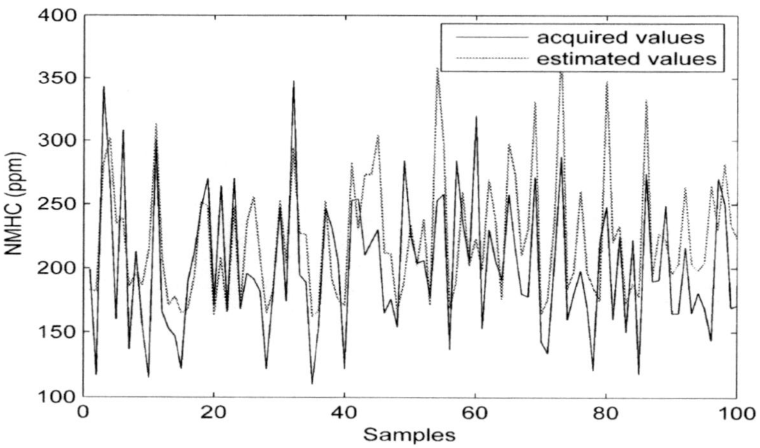

Figure 8.13. MNHC (in ppm) levels estimated using the neural network based soft sensors for the removed station

Similar results were obtained for the soft sensors used to model the other removed station.

The two removed stations became available for reallocation in the urban area. By using the spline method, a family of two-dimensional static maps for each pollutant was determined. Each map in the family allows one to estimate the values of pollutant concentration in continuous spatial coordinates in the urban area, at a given time. By analyzing the map family with respect to spatial and time coordinates, points corresponding to the maximum level for at least one pollutant were considered as candidate points for reallocation.

As an example, a map of the CO pollutant level at a given time is illustrated in Figure 8.14. Points A and B correspond to extreme values of the CO level and are therefore possible points for the reallocation of monitoring stations.

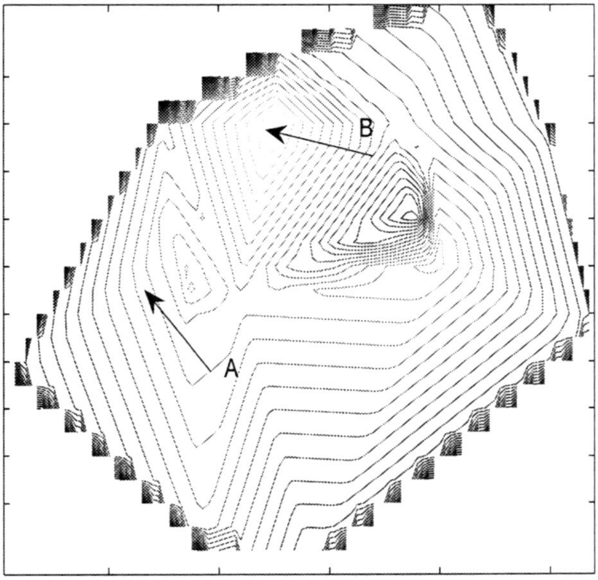

Figure 8.14. Two-dimensional map of the CO level used for the reallocation of an air pollution monitoring station

The application reported in this section is a valid example of the potential for soft sensors to be used in areas other than industrial environments. Also, in the reported study, a distributed model was required.

8.4 Conclusions

The cases of study reported in this chapter have been introduced with the aim to give to the reader a wider view of possible applications of soft sensors. In fact, in previous chapters we focused on monitoring of industrial plants. Here two different applications have been reported.

In the first case a dynamic data-driven model was described, designed to perform the what-if analysis of an industrial plant. This type of applications suggests the possibility to exploit soft sensors as a tool for the design of control policies. It is mandatory for this kind of applications to be very restrictive in assessing the soft sensor quality. This requires a careful attention in data selection, during the design phase, to assure that relevant dynamics are not missed. Also, along with the usual model validation step, when possible *ad hoc* measuring surveys should be performed on the plant by using the a representative subset of input trends given to the soft sensor.

In the second case a set of soft sensors were described that were designed for monitoring purposes of air pollution in large urban areas. This shows that soft sensors can profitably be used in applications fields different from industrial processes.

The described application shows that soft sensors have interesting application possibilities for the management of distributed monitoring networks. In fact, soft sensors can be used to optimize used hardware location and to interpolate data in points where measuring hardware is not available, thus alleviating the drawbacks due to the discrete nature of the hardware monitoring network.

9

Fault Detection, Sensor Validation and Diagnosis

9.1 Historical Background

Since the early 1960s, industrial processes have undergone a huge increase in their degree of automation, due to increased demands on performance and the need to release human operators from constant attendance at the plant. As a consequence, the need for adequate tools for automatic fault detection, diagnosis and supervision has been strongly felt since then.

In the industrial application context, a fault is to be understood as a non-permitted deviation of a characteristic property, which leads to the inability of a system or a plant to fulfil the purpose it has been designed for (Isermann, 1997). A certain amount of effort has been devoted by the scientific community to establish a common terminology (Isermann and Ballé, 1997; Omdahl, 1988) which is reported at the end of this chapter. Nevertheless, the multidisciplinary peculiarity of the topics involved often leads to a terminology which is not unique.

The development of automatic fault detection strategies took place in the early 1970s. The first strategies were based on linear observers, operating on linear systems. A survey of these techniques is reported in Willsky (1976). One of the first books on the subject was published in 1978, about fault detection and diagnosis in chemical and petrochemical processes (Himmelblau, 1978). A lot of work has been carried out using the so-called analytical redundancy paradigm, exploiting well-established automatic control techniques like parameter and state estimation. These techniques are summarized in Isermann (1984). In the same period, parity-equation-based strategies were treated (Patton and Chen, 1991; Gertler, 1991; Hoefling and Pfeufer, 1994). The state of the art in the 1980s is summarized in books (Pau, 1981; Patton, Frank, and Clark, 1989), and survey papers (Gertler, 1988; Frank, 1990; Isermann, 1994). Many books provide a clear idea of the developments in the field (Patton, Frank, and Clark, 2000; Poulizeus and Stravlakakis, 1994; Gertler, 1998; Chen and Patton, 1999; Chiang, Russel and Braatz, 2001; Simani, Fantuzzi, and Patton, 2002; Isermann, 2006a; Isermann, 2006b). With the rising interest in fault detection, a certain number of engineering associations created workgroups to deal with novel, emerging issues. As an example, IFAC (International Federation of Automatic Controls) created in 1991 the SAFEPROCESS (fault detection, supervision, and safety for technical processes) Steering Committee, which then became an IFAC Technical Committee in 1993. An IFAC SAFEPROCESS Symposium is regularly scheduled every three years (IFAC, 1991–2006), together with the IFAC workshop for fault detection in supervision in chemical industries (IFAC-Workshop, 1986–2001).

9.2 An Overview of Fault Detection and Diagnosis

A fault can occur in any of the components of an industrial process. With respect to the location of a fault, one can distinguish among sensor faults, actuator faults, and component or process faults. When a fault occurs, a chain of actions can be undertaken in order to cope with it. These actions can be ordered in a hierarchical fashion.

The first action when a fault occurs is fault detection, which basically consists of revealing the presence of a fault, possibly revealing also the time of the faulty event.

Once the fault is detected, the second step is fault isolation, which consists in determining the kind, location, and time of the fault. Subsequently, fault identification must be performed in order to determine the size and time-variant behavior.

The complex of fault detection, isolation, and identification is called fault diagnosis. Some authors refer instead to the term fault detection and diagnosis (FDD), thus considering the detection as a separate task, and including only isolation and identification in the diagnosis activity. In a broader sense (and maybe closer to the commonly held concept), a diagnosis can be seen as an attempt to explain system misbehavior by analyzing the relevant features that characterize it (Ulieru and Mrsic-Flogel, 1994). These characteristics are often called symptoms, or sometimes fault indicators. Essentially, diagnosis is always related with symptom observation (clearly, this diagnosis process is better-known and established in the case of human diseases rather than in industrial faults).

A more complex step to perform is fault evaluation, the aim of which is to provide an estimate of how the detected and diagnosed fault will affect the future life of the process.

On the basis of the results achieved through one or more of the steps described above, decisions have to be made, either by human experts or in an automated fashion. In other words, the fault must be managed. Usually, the most common decisions are:

- the fault is tolerable: the system can continue its operations;
- the system has to change its operational scenario in order to cope with the fault (through reconfiguration, maintenance, or repair);
- the system has to be stopped and the fault eliminated.

One of the most important issues in the whole fault detection and diagnosis process is the concept of redundancy. This is the capability of having two or more ways to determine some characteristic properties (variables, parameters, signal features, *etc.*) of the process, in order to exploit more information sources for performing an effective detection and diagnosis action. Whichever fault detection technique one wishes to implement, redundancy is the fundamental paradigm to exploit.

The most direct form of redundancy is physical redundancy. This consists of physically replicating components of the system, either hardware or software modules. Physical redundancy can be a way to achieve fault-tolerant systems, *i.e.* systems which are able to continue to work in a proper way, despite the presence of faults, by properly switching the modules affected by faults with spare, back-up models. Duplicating modules can also be a first approach to fault detection. If all replicated modules behave in the same way, it can be concluded that no one is affected by a fault. Comparing the behavior of replicated modules is therefore a first strategy to detect a fault. This technique is often used with sensor validation (Dorr *et al.*, 1997), which will be dealt with in detail later in this chapter.

When physical and mathematical relationships among variables of a system are known, and causes and effects can then be detected and isolated, physical redundancy can be overcome by the presence of another redundant source of information, which is called analytical redundancy. In other words, a model of the physical phenomena involved (or parts of them) can be built. As stated in previous chapters, the process of building a model is usually twofold.

On the one hand, models can be constructed on the basis of physical and chemical laws which describe the system behavior. This approach, also known as mechanistic modeling, involves the formulation of assumptions about the nature of the system, its causes (inputs) and their related effects (outputs), and the establishing of a mathematical description of the system behavior. Very often some simplifying assumptions about the process are made in order to reduce the complexity of the model, without any significant detriment to the reproduction of the system behavior. Moreover, the system is often divided into control volumes, connected by suitable connectors (which can be either physical or logical), through which the subsystems exchange masses, flows, energy, information, and so on. This is called the Eulerian approach.

The physical laws involved in the description of the system behavior are usually the following:

1. balance equations for masses, energies and impulses;
2. constitutive physical–chemical state equations;
3. phenomenological and/or entropy balance equations for irreversible processes;
4. connection equations, describing the interconnection of the modeled elements.

This approach leads to a description of the system behavior in terms of differential equations, depending on certain (known or unknown) parameters. The complexity of the equations involved (from low-order, linear, time-invariant, ordinary differential equations to high-order, nonlinear, time-variant, partial differential equations) depends on the nature of the system under analysis and on the simplifying assumptions made.

On the other hand, one can exploit the availability of process measurements to construct models based on the usual identification procedures.

In practical applications, models can often be derived from a fusion of physical knowledge, identification, and heuristic practice. Identification and heuristics may be more and more pervasive as long as constitutive laws are unknown or complex, as is illustrated schematically in Figure 9.1.

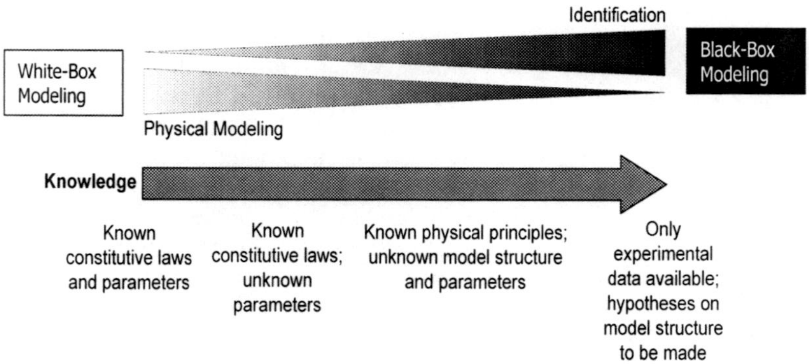

Figure 9.1. Concurrence of both identification and physical modeling in practical applications

When a model of a component is used for fault detection purposes, it is fed with the same inputs as the actual component. In normal working conditions, the output of the model and that of the actual system should coincide. Therefore, an indicator of fault emerges from the comparison between actual and model outputs, *i.e.* the model residual. Residual analysis is the core of all model-based fault detection and diagnosis techniques. Model-based fault detection will be dealt with in depth in the next section.

Perhaps the most fascinating (but complex to manage) form of redundancy is knowledge redundancy. This derives from: human experience; the results of past experiments, faults, repairs; the heuristic knowledge of the phenomena, in terms of peculiar noise, colors, smells, and other vague information (including vague values for variables); human ability to connect apparently unrelated symptoms and causes.

Knowledge redundancy is appealing and effective, but needs specific computational tools to be integrated in fault detection and diagnosis schemes based on mathematical and/or statistical approaches. With this in view, the well-established soft computing techniques (Fortuna *et al.*, 2001), are among the most effective means to integrate all the kinds of redundancy, including human knowledge, heuristics, and past experience, to cope with complex problems, taking into account the undetermined and imprecise nature of the real world.

9.3 Model-based Fault Detection

As previously stated, when a model of the process is available, fault detection and diagnosis can be performed by comparing the behavior of the model with that of the component under analysis. Discrepancies between the behavior of the process and the model indicate that a fault has occurred. This is called the model-based approach; the form of redundancy emerging from this approach is called analytical redundancy.

The appealing aspect of analytical redundancy relies in the fact that redundancy is achieved by a mathematical treatment of available information, rather than through the physical replication of components. Obviously, this benefit is offset by the cost and effort needed to obtain a reliable mathematical model of the process. One of the main issues related to model-based fault detection is the sensitivity of the detection system with respect to modeling errors, which are by no means avoidable in practice. It seems clear that modeling errors can actually obscure the effects of faults or exaggerate slight deviations from the nominal system behavior, thus being potential sources of false alarms.

For an effective implementation of a model-based fault detection and isolation (FDI) system, one has to take into account all the causes that can lead either to alarms or false alarms, in particular:

- faults in the actuators, components of the plant, or sensors;
- modeling errors;
- system noise and measurement noise.

A simple conceptual scheme for model-based fault detection is reported in Figure 9.2 (Frank, 1990).

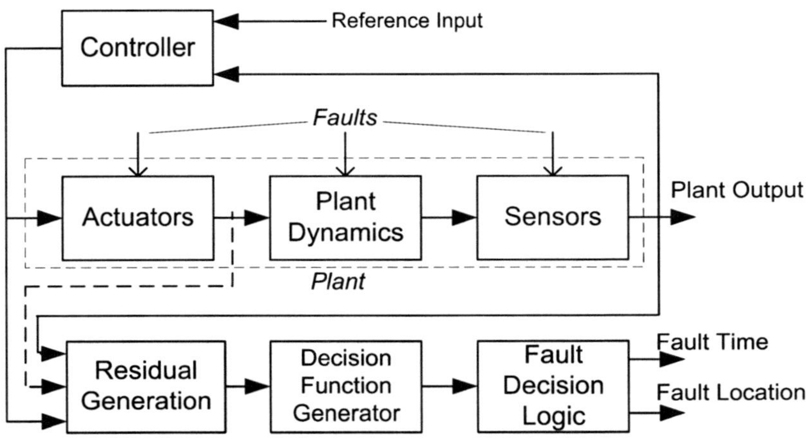

Figure 9.2. Conceptual scheme for model-based fault detection (Frank, 1990)

9.3.1 Fault Models

Different kinds of faults can occur in a system. Like systems, faults can also be mathematically modeled. Modeling faults can be useful to develop effective FDD tools, especially when they are based on mathematical modeling of the system. In fact, once the model of the fault-free system has been developed, determining mathematical models for most common faults can lead to the development of mathematical models for the faulty system. This is done by studying the effects of the mathematically modeled faults on the system model equations.

The mathematical modeling of faults is done by studying the actual faulty system. Faults can arise from many causes; for example, errors made in design or assembly, wrong operation, lack of maintenance, corrosion, aging, wear during normal operational conditions, or external agents, like noise, disturbance, human errors, *etc.*

Moreover, they can appear suddenly or gradually, with small or large size. Faults which keep acting on the system once they occur are usually called deterministic faults, and they are modeled through deterministic changes in the system or signal equations. On the other hand, intermittent faults (which are harder to detect) are called stochastic faults, and their modeling must generally be performed within a stochastic framework.

A fault is an unpermitted deviation of a characteristic property of the system, which can be any physical quantity. Let us assume that this quantity can be described by a mathematical law

$$Y(t) = g[U(t), \mathbf{x}(t), \boldsymbol{\theta}] \tag{9.1}$$

where $Y(t)$, $U(t)$, $\mathbf{x}(t)$, $\boldsymbol{\theta}$, are the output variable, input variable, state variable vector, and parameter vector, respectively. Faults can therefore occur as changes in signals or parameters.

With respect to the time of appearance of the faults, the following kinds of faults can be distinguished:

- abrupt fault (stepwise);
- incipient fault (drift);
- intermittent fault.

With respect to the way of influencing the equations describing the time evolution of the physical quantity, faults can be classified in the following main classes:

- additive faults;
- multiplicative faults.

An additive fault modifies the quantity $Y(t)$ by the addition of a quantity $f(t)$, representative of the effect of the fault

$$Y(t) = Y_u(t) + f(t) \tag{9.2}$$

where $Y_u(t)$ represents the nominal time evolution of $Y(t)$ under the effect of the input $U(t)$.

Multiplicative faults modify nominal quantities by multiplication by a quantity $f(t)$. For simplicity, let us assume that the nominal relation between $Y(t)$ and $U(t)$ is given by $Y_u(t)=aU(t)$, where a is a system parameter. In this case, the fault $f(t)$ affects $Y(t)$ as

$$Y(t) = (a + \Delta a(t))U(t) = Y_u(t) + f(t)U(t) \tag{9.3}$$

Faults can also affect the input signal, or a state variable, instead of the output signal.

In order to detect the change $\Delta Y(t)$ caused by the presence of the fault, it must be noted that in the case of additive faults this change is independent of any other signal

$$\Delta Y(t) = f(t) \tag{9.4}$$

whereas, for multiplicative faults, the change $\Delta Y(t)$ depends on other signals. For example, considering Equation 9.3, we obtain

$$\Delta Y(t) = f(t)U(t) \tag{9.5}$$

This implies that the change in $Y(t)$ depends on the size of $U(t)$. Moreover, in the case of a multiplicative fault, it can be detected only if $U(t)\neq 0$.

9.3.2 Fault Detection Approaches

In the model-based framework, once residuals are generated, they must be evaluated and a decision must be made, in order to detect the fault. In the simplest form, residuals are generated by comparing output variables of the model against the corresponding output variables of the system. Several techniques inspired by system theory can be applied to generate residuals and perform fault detection. In particular:

- parity checks. The key idea of this strategy consists of checking the parity (consistency) of the mathematical equations of the system by exploiting the actual measurements (Gertler, 1997; Gertler, 1998);
- observer-based schemes (Patton and Chen, 1997; Frank, 1990). This technique consists of reconstructing the state or the output of the system from the available measurements, by means of Luenberger observers (in the deterministic case) or Kalman filters (in the stochastic case, when noise must be taken into account), using the estimation error or the innovation, respectively, as fault indicators (residuals);
- parameter estimation techniques (Isermann, 1984; Frank, 1990). These techniques rely on the fact that the occurrence of a fault is often reflected in one or more physical parameters of the system. If this is the case, some

parameters of the process can be continuously estimated, and residuals can be generated by comparing the estimated parameters with the nominal ones, which had been estimated during a fault-free condition. Consequently, residuals are generated by comparing estimated and nominal (fault-free) parameters.

9.3.2.1 Fault Detection with Parity Equations

The main idea underlying the parity equation approach is to exploit a model of the non-faulty system in order to compare its output with the actual output of the system. Let us refer to Figure 9.3, which shows two possible parity-equation-based fault detection schemes for linear processes, described in terms of transfer functions. For the sake of simplicity, let us consider SISO processes.

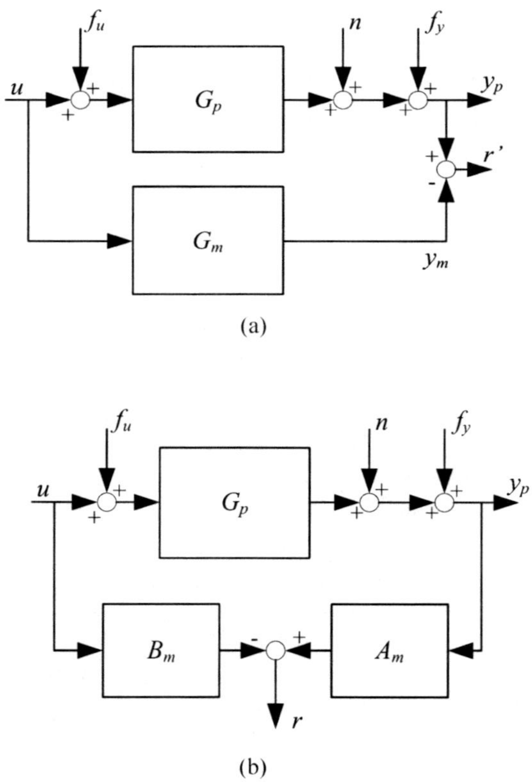

Figure 9.3. Residual generation with parity equation schemes. (a) Output error, (b) polynomial error (or equation error)

The process under analysis, with input u and output y_p, is affected by two additive faults, f_u and f_y, acting on the input and the output, respectively. Moreover, an additive noise, n, acts on the output. Process and model are described by the two transfer functions, $G_p(s)$ and $G_m(s)$, respectively

$$G_p(s) = \frac{y_p(s)}{u(s)} = \frac{B_p(s)}{A_p(s)} \tag{9.6}$$

$$G_m(s) = \frac{y_m(s)}{u(s)} = \frac{B_m(s)}{A_m(s)} \tag{9.7}$$

The transfer function of the process is unknown, whereas that of the model is obviously known. Therefore, the two transfer functions are related by means of a transfer function describing modeling errors, $\Delta G_m(s)$

$$G_p(s) = G_m(s) + \Delta G_m(s) \tag{9.8}$$

A first approach to define residuals is to compute the output error, that is the difference between process and model output, as illustrated in Figure 9.3(a).

$$\begin{aligned} r'(s) &= y_p(s) - y_m(s) = y_p(s) - G_m(s)u(s) \\ &= G_p(s)[u(s) + f_u(s)] + n(s) + f_y(s) - G_m(s)u(s) \\ &= \Delta G_m(s)u(s) + G_p(s)f_u(s) + n(s) + f_y(s) \end{aligned} \tag{9.9}$$

If neither faults, nor noise, are present, the residual in Equation 9.9 is zero. A residual value different from zero may be caused by modeling errors, noise, or faults. The effect of faults f_y on the output are directly reflected onto the residual changes, whereas faults on the input f_u influence the residual changes after being filtered by the process G_p. Moreover, the effect of modeling errors on the residuals depends on the input signal, u, and noise is directly propagated, by addition, onto the residual.

An alternative method to define residuals is through the polynomial error (or equation error), as illustrated in Figure 9.3(b):

$$\begin{aligned} r(s) &= A_m(s)y_p(s) - B_m(s)u(s) = \\ &= A_m(s)\big[G_p(s)[u(s) + f_u(s)] + n(s) + f_y(s)\big] - B_m(s)u(s) \end{aligned} \tag{9.10}$$

If the model is an exact representation of the process (*i.e.* $\Delta G_m(s)=0$), the residual is

$$r(s) = A_m(s)\big(f_y(s) + n(s)\big) + B_m(s)f_u(s) \tag{9.11}$$

Also in this case, the residual is affected by both input and output faults, and by the noise. Output faults and noise are filtered by the polynomial $A_m(s)$, whereas input faults are filtered by the polynomial $B_m(s)$. Unlike in the output error approach, the filtering polynomial involved in the equation error approach can lead to problems related to realizability and to amplification of the noise, especially if its spectrum is located at high frequencies.

Equations 9.9 and 9.10 are called parity equations, and r and r' are called primary residuals (Gertler, 1998).

A drawback of the parity equation approach is that primary residuals are affected by input and output faults, by the noise, and by the modeling errors. Usually, the effects of these causes are hardly separable. A slight improvement can be achieved if more output measurements are available, which is the case for MIMO systems. The approach can easily be extended by considering adequate transfer function matrices (Gertler, 1998). Consequently, the residuals in Equations 9.9 and 9.10 become residual vectors, with the same size as the output vector. In this case, it may happen that different kinds of faults affect different elements of the residual vector, leading to a separation of the possible faults.

The parity equation approach can be reformulated within different frameworks, such as input–output models (Gertler and Singer, 1985), state-space (Chow and Willsky, 1984; Gertler, 1991), and enhanced state-space models (Patton and Chen, 1991).

Concerning nonlinear systems, because of the vast amount of possible nonlinear processes and models, there are no general approaches to the problem of generating parity equations. The way of generating residuals is strongly related with the structure adopted for the nonlinear model.

If the nonlinear model can be described by

$$y_m(t) = g\left[\dot{y}(t), \ddot{y}(t), \ldots; u(t), \dot{u}(t), \ddot{u}(t), \ldots\right] \tag{9.12}$$

we can consider output residuals such as

$$r(t) = y_p(t) - y_m(t) \tag{9.13}$$

where $y_p(t)$ is the process output. In the nonlinear case, the study of the effect of faults, noise, and modeling error on residual changes is more complicated than in the linear case.

The process can also be described by discrete-time polynomial models like Hammerstein or parametric Volterra models (Eykhoff, 1974; Haber and Unbehauhen, 1990), and output and equation errors can be correspondingly defined. Also in this case, the study of the effects of faults, noise and modeling errors on residual changes is not straightforward.

Alternatively, artificial neural networks, fuzzy or neuro-fuzzy models can be adopted to model nonlinear processes (Chen, Billings and Grant, 1990; Takagi and Sugeno, 1985; Fortuna *et al.*, 2001).

9.3.2.2 Residual Properties

As stated above, in the parity equation approach residual changes are caused not only by faults, but also by modeling errors, noise, unknown inputs and disturbances. Clearly, one might wish, instead, that residual changes were only caused by faults.

As the ideal condition cannot be achieved (*i.e.* the residual is identically equal to zero if faults are not present), an effective fault detection scheme requires also that thresholds for residual checks are fixed. The choice of a threshold always comes from a trade-off between accuracy and robustness of the detection scheme, as greater thresholds lead to fewer false alarms, but do not allow us to detect small faults.

Several techniques can be adopted to generate residuals which convey better information about faults, thus allowing the designer to choose smaller thresholds without affecting robustness:

1. adequate low-pass filtering for high-frequency disturbances;
2. robustness against modeling errors;
3. adaptive thresholds;
4. enhanced residuals for specific faults;
5. enhanced sensitivity of residuals with respect to faults.

The first three items of the list can be achieved using well-known signal processing or system modeling techniques. The main idea underlying the generation of enhanced residuals is to design the parity checks such that the residual vector has specific properties which characterize the residual according to the fault which has occurred. In this way, faults may be distinguished from each other. Two main approaches to generate enhanced residuals are the generation of structured and directional residuals (Gertler, 1998).

Structured residuals have the property that only some elements of the residual vector are different from zero when a specific fault occurs. This means that each kind of fault causes changes only in certain subspaces in the vector space spanned by the residual vector. This implies that different residual patterns emerge from the residual vector in correspondence with different faults. The residual patterns are often called fault signatures.

On the other hand, directional residuals are designed to make the residual vector assume a given direction in vector space under the occurrence of a specific fault. Therefore, the direction of the residual vector indicates the type of fault. Moreover, the modulus of the residual vector is an indicator of the fault size.

Concerning the last item of the list, the enhancement of the sensitivity of residuals with respect to faults is treated within the state-space approach through a typical sensitivity analysis of the analytical formulation of the residual vector. The topic is well addressed in Hoefling and Isermann (1996).

9.3.2.3 Fault Detection with Observer-based Schemes

As stated above, observer-based fault detection exploits the observer estimation error as fault indicator (residual). In order to show the basic technique, let us focus our attention on a linear time-invariant MIMO system, described by the state space model

$$\dot{\mathbf{x}}(t) = \mathbf{A}\mathbf{x}(t) + \mathbf{B}\mathbf{u}(t) \tag{9.14}$$

$$\mathbf{y}(t) = \mathbf{C}\mathbf{x}(t) \tag{9.15}$$

where $\mathbf{u}(t)$ is the input vector, $\mathbf{y}(t)$ is the output vector, and $\mathbf{x}(t)$ is the state vector. It is well known that, if the system is observable, a full-order observer can be designed. Observer inputs are input and output measurements of the system ($\mathbf{u}(t)$ and $\mathbf{y}(t)$); observer output is an estimate $\hat{\mathbf{x}}(t)$ of the state $\mathbf{x}(t)$. The observer dynamics is governed by the equations

$$\dot{\hat{\mathbf{x}}}(t) = (\mathbf{A} - \mathbf{H}\mathbf{C})\hat{\mathbf{x}}(t) + \mathbf{B}\mathbf{u}(t) + \mathbf{H}\mathbf{y}(t) \tag{9.16}$$

$$\hat{\mathbf{y}}(t) = \mathbf{C}\hat{\mathbf{x}}(t) \tag{9.17}$$

where $\mathbf{H}$ is the observer feedback gain matrix, which must be properly chosen in order to achieve the desired observer performance. Under the assumption that process and model parameters are exactly known, the estimation error

$$\tilde{\mathbf{x}}(t) = \mathbf{x}(t) - \hat{\mathbf{x}}(t) \tag{9.18}$$

is governed by

$$\dot{\tilde{\mathbf{x}}}(t) = [\mathbf{A} - \mathbf{H}\mathbf{C}]\tilde{\mathbf{x}}(t) \tag{9.19}$$

Consequently, the estimation error tends asymptotically to zero for any initial state deviation $[\mathbf{x}(0) - \hat{\mathbf{x}}(0)]$ if $[\mathbf{A} - \mathbf{H}\mathbf{C}]$ is asymptotically stable. This can always be achieved by pole placement with a suitable choice of the matrix $\mathbf{H}$, if and only if the couple $(\mathbf{A}, \mathbf{C})$ is observable.

Let us now assume that the process is affected by unmeasurable state and output disturbances, $\mathbf{v}(t)$ and $\mathbf{n}(t)$, and by additive state and output faults, $\mathbf{f}_l(t)$ and $\mathbf{f}_m(t)$, as follows

$$\dot{\mathbf{x}}(t) = \mathbf{A}\mathbf{x}(t) + \mathbf{B}\mathbf{u}(t) + \mathbf{V}\mathbf{v}(t) + \mathbf{L}\mathbf{f}_l(t) \tag{9.20}$$

$$\mathbf{y}(t) = \mathbf{C}\mathbf{x}(t) + \mathbf{N}\mathbf{n}(t) + \mathbf{M}\mathbf{f}_m(t) \tag{9.21}$$

where $\mathbf{L}$, $\mathbf{M}$, $\mathbf{V}$, $\mathbf{N}$ are fault and disturbance entry matrices. Application of the full-state observer expressed in Equation 9.16 to the model described by Equations 9.20 and 9.21 leads to the reconstruction error

$$\dot{\hat{\mathbf{x}}}(t) = (\mathbf{A} - \mathbf{H}\mathbf{C})\hat{\mathbf{x}}(t) + \mathbf{V}\mathbf{v}(t) + \mathbf{L}\mathbf{f}_l(t) - \mathbf{H}\mathbf{N}\mathbf{n}(t) - \mathbf{H}\mathbf{M}\mathbf{f}_m(t) \tag{9.22}$$

and to the output error

$$\mathbf{e}(t) = \mathbf{y}(t) - \mathbf{C}\hat{\mathbf{x}}(t) = \mathbf{C}\hat{\mathbf{x}}(t) + \mathbf{N}\mathbf{n}(t) + \mathbf{M}\mathbf{f}_m(t) \tag{9.23}$$

Observing Equations 9.22 and 9.23, it is clear that, after a suitable time interval such that the initial state deviation $[\mathbf{x}(0) - \hat{\mathbf{x}}(0)]$ vanishes, the estimation error $\tilde{\mathbf{x}}(t)$ and the output error $\mathbf{e}(t)$ are affected both by disturbances and faults. In particular, after the asymptotic vanishing of the initial state deviation, the estimation and output errors are zero if neither faults nor disturbances are present, and are different from zero if faults or disturbances are present. Therefore, the estimation and output errors can be considered as suitable residuals for fault detection purposes.

If the input–output relation of the state observer described by Equation 9.16 is derived, an interesting similarity with the parity equation approach can be noticed: while in the output error approach based on parity equations, faults on the input influence the residual changes after being filtered by the process transfer function $G_p(s)$, in the observer-based scheme they are filtered by the observer dynamics $[s\mathbf{I} - (\mathbf{A} - \mathbf{HC})]^{-1}$.

For multiplicative faults, performing fault detection is more complicated. As an example, let us consider changes $\Delta\mathbf{A}$, $\Delta\mathbf{B}$, $\Delta\mathbf{C}$ in the parameters of the state-space model. Consequently, the system model is

$$\dot{\mathbf{x}}(t) = [\mathbf{A} + \Delta\mathbf{A}]\,\mathbf{x}(t) + [\mathbf{B} + \Delta\mathbf{B}]\,\mathbf{u}(t) \tag{9.24}$$

$$\mathbf{y}(t) = [\mathbf{C} + \Delta\mathbf{C}]\,\mathbf{x}(t) \tag{9.25}$$

and, neglecting disturbances, the state and output error equations become

$$\dot{\tilde{\mathbf{x}}}(t) = [\mathbf{A} - \mathbf{HC}]\,\tilde{\mathbf{x}}(t) + [\Delta\mathbf{A} - \mathbf{H}\Delta\mathbf{C}]\,\mathbf{x}(t) + \Delta\mathbf{B}\mathbf{u}(t) \tag{9.26}$$

$$\mathbf{e}(t) = \mathbf{C}\tilde{\mathbf{x}}(t) + \Delta\mathbf{C}\mathbf{x}(t) \tag{9.27}$$

In Equations 9.26 and 9.27 it can be clearly noticed that faults are reflected onto the residuals after being multiplied by the input signal and by the state variables. This makes the analysis of multiplicative faults more complicated.

The introductory material presented so far about observer-based fault detection schemes illustrates the working principle of the approach. In the design of an observer for fault detection, the main design issue is to optimize the choice of the feedback gain matrix $\mathbf{H}$ in order to achieve good fault detection capability. Many different observer-based approaches have been developed, including reduced-order observers, observer banks, dedicated observer schemes, Kalman filters, and so on.

Some approaches are of particular interest for fault detection as they allow the concept of structured and directional residuals to be incorporated into the observer-based framework. This is the case of the fault detection filter (FDF) approach, which relies on a particular choice of the feedback gain matrix $\mathbf{H}$ such that the residual due to a particular fault is constrained to a single direction or plane in the residual space. To achieve this goal, a state augmentation of the system and observer model is often needed. State augmentation is a well-known technique in which some system parameters are interpreted as further state variables, thus increasing the order of the system. Consequently, if the augmented state is observed by an (augmented) observer, the fault occurring in a particular parameter can be detected and isolated as if it were detected in a state variable. For an extensive survey on observer-based fault detection techniques, refer to Chen and Patton (1999), Patton and Chen (1997), and Frank (1990).

9.3.2.4 Fault Detection with Parameter Estimation

The two techniques described above belong to the framework of signal estimation techniques. Alternatively, parameter estimation techniques may be adopted for fault detection purposes. Fault detection with parameter estimation relies on the fact that very often faults in a dynamic systems have an effect on the value of physical parameters, like for example mass, friction, electric resistance, *etc.*

The approach is based on comparison of the estimated (actual) parameters with the nominal ones. In what follows, we outline the procedure for fault detection based on parameter estimation (Isermann, 1984), leaving aside the specific techniques for parameter estimation, as they are part of well-established identification techniques (Ljung, 1999):

1. Choice of a suitable parametric model of the system. The simplest choice that can be made is a linear, lumped parameter, input/output differential equation of the form:

$$\sum_{i=0}^{n} a_i \frac{d^i y(t)}{dt^i} = \sum_{j=0}^{m} b_j \frac{d^j u(t)}{dt^j} \tag{9.28}$$

2. As parameter identification techniques usually rely on black-box modeling, physical parameters p do not always coincide with model parameters a_i, b_j. Therefore, a relationship between physical parameters and model parameters must be established as

$$\boldsymbol{\theta} = \begin{bmatrix} a \\ b \end{bmatrix} = \mathbf{f}(\mathbf{p}) \tag{9.29}$$

3. Identification of the model parameter vector $\boldsymbol{\theta}$ on the basis of available input/output measurement by means of a suitable identification technique (Ljung, 1999).
4. Determination of the physical parameter vector as

$$\mathbf{p} = \mathbf{f}^{-1}(\boldsymbol{\theta}) \tag{9.30}$$

5. Comparison of the estimated parameter vector **p** with the nominal parameter vector **p*** worked out from the nominal (fault-free) system or model. Calculation of the fault indicator

$$\Delta \mathbf{p} = \mathbf{p} - \mathbf{p}^* \tag{9.31}$$

6. Detection of the occurrence of a fault on the basis of the fault indicator $\Delta\mathbf{p}$ and the known relationships between faults and parameter changes.

9.3.3 Improved Model-based Fault Detection Schemes

A more detailed conceptual scheme for fault detection based on analytical redundancy is shown in Figure 9.4. It is important to notice that for effective fault detection and isolation, three models are often needed (Isermann, 1984), namely the model of the nominal system, the model of the actual (observed) system, and that of the faulty system. Moreover, in order to keep the performance of the detection and isolation at an acceptable level, the nominal model must itself be monitored, and updated by novel information about the actual system coming from the observation model (*i.e.* in the case in which the system undergoes an important change in its operational state, without entering a faulty state, the nominal model must be updated).

As previously stated, inaccuracies in modeling may affect the performance of fault detection algorithms. In fact, due to modeling inaccuracies, nonzero residuals may be generated without the occurrence of a fault. This is in practice unavoidable. Therefore, residuals must be evaluated more thoroughly, through the adoption of suitable thresholds or other accurate checks. It is obvious that the design of a fault detection and isolation system involves a trade-off between the accuracy of the fault detection and the generation of false alarms. An important issue in the design of fault detection systems is the reduction of the residuals sensitivity with respect to modeling uncertainties. Strategies devoted to the improvement of robustness in model-based fault detection systems are reported in Patton, Frank and Clark (1989), Chow and Willsky (1984), Lou, Willsky and Verghese (1986), Patton and Chen (1997). The issues related to a proper choice of thresholds on the residuals, also providing adaptation, are reported in Emami-Naeini (1986), Patton, Frank and Clark (1989).

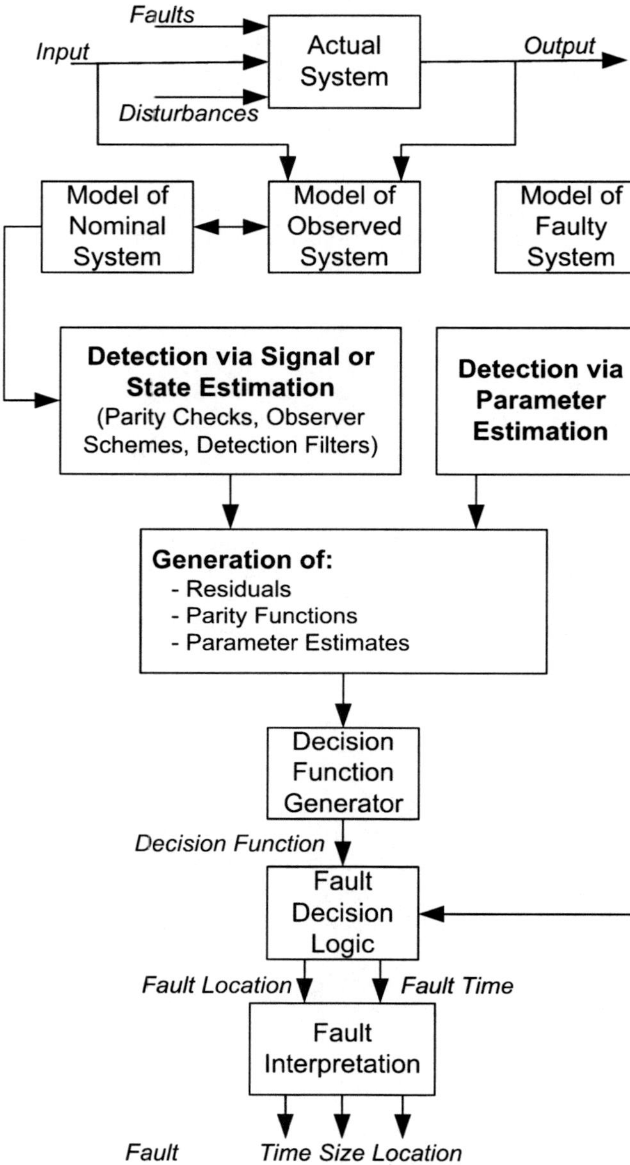

Figure 9.4. General scheme for model-based fault detection and diagnosis (Frank, 1990)

9.4 Symptom Analysis and Fault Diagnosis

After a fault has been detected, the next step to perform is diagnosis. The aim of diagnosis is to provide as much information as possible about the fault that has occurred (or is occurring), like time of detection, location and size of the fault. Fault diagnosis implies the extraction of suitable features about the status of the system to be monitored. These may come from several sources of redundancy (physical, analytical, knowledge), being, for example, estimated parameters or state variables, residuals, *etc.*

Subsequently, features are compared with their nominal value. If feature changes are detected, a fault is detected and the feature change is called a symptom. Therefore, symptoms are information which reflects changes from normal system behavior. The first kind of symptoms that can be considered are the residuals generated within a model-based approach, but the field is not limited to this. Effective fault detection and diagnosis strategies may often involve both mathematical processing of variables and their thorough evaluation by skilled personnel with a good heuristic knowledge. This often leads one to consider fault detection and diagnosis within a knowledge-based approach (Rasmussen, 1993; Struss, Malik and Sachenbacher, 1996; Isermann, 1994; Isermann, 1997). From a more general point of view, symptoms can be classified into three main classes (Isermann, 1997; Isermann, 2006a):

- analytic symptoms;
- heuristic symptoms;
- symptoms emerging from process history or fault statistics.

Analytic symptoms are generated by exploiting the analytical knowledge about the system under analysis, in order to generate quantifiable, analytical information. In other words, analytic symptoms can be generated when data processing are performed on measurable process quantities. Consequently, characteristic values able to indicate the presence and possibly the nature of the fault are generated. These characteristic values are mainly generated through three techniques:

- process analysis using mathematical models of the process, through the use of state estimation, parity checks or parameter estimation;
- signal analysis of directly measurable signals by exploiting suitable signal models, like correlation functions, frequency spectra, ARMA signal models, and so on;
- limit checking performed on directly measurable quantities of the process. The symptoms are the signal tolerances exceeded.

Heuristic symptoms can be generated in addition to analytic ones, by taking into account qualitative information coming from human experience. For example, heuristic knowledge about the process in terms of changes in colors, smells, special noises, vibrations, wear and tear, and so on, can be considered. The obvious difficulty in taking into account heuristic symptoms is their formalization in a machine-friendly format. Heuristic symptoms are usually expressed in terms of linguistic variables, or as vague quantitative information, typically around a certain nominal value with undefined tolerance. For this kind of symptoms, one of the most used paradigms is fuzzy logic (Fortuna *et al.*, 2001; Isermann and Ulieru, 1993; Ulieru and Mrsic-Flogel, 1994; Dexter and Benouarets, 1997; Ballé, 1998).

Process history and fault statistics can be exploited in order to generate symptoms which take into account the past life of the process. This can include, for example, information on running times, load measures, maintenance, and repairing activity. If statistics on faults have been performed, information concerning the frequency and location of specific faults can be generated, in order to evaluate symptoms based on these statistics. Actually, this third category of symptoms can fall either in the analytical class or in the heuristic one, depending on the accuracy and reliability of measurements and statistics. However, generally, the information collected through process history is rather vague and inaccurate, and is usually taken into account in terms of heuristics information.

In the field of fault diagnosis and classification, soft computing techniques (Fortuna *et al.*, 2001) can be extremely useful in integrating all kinds of symptoms in a unified fashion. For example, a fuzzy representation of the symptoms is suitable for both heuristic and analytical symptoms. In fact, it allows the designer to describe heuristic symptoms in terms of vague, linguistic concepts; and to incorporate the unavoidable uncertainties into the description of analytic symptoms.

Once symptoms are generated, the association between the occurrence of a specific fault and its related symptoms must be identified.

If no or little information about this association is available, classification or pattern recognition approaches can be suitable to cope with this problem. Let us consider a system in which m different faults can occur, and n different symptoms can be generated. Symptoms can be collected in a symptom vector

$$S = [S_1 \quad S_2 \quad \cdots \quad S_n] \tag{9.32}$$

and the faults occurring can be considered as a fault vector, indicating the occurrence of faults

$$F = [F_1 \quad F_2 \quad \cdots \quad F_m] \tag{9.33}$$

The generic element of the fault vector F can be considered as either a binary variable ($F_j \in [0,1]$), in which the value 1 indicates the occurrence of a fault; or as a continuous variable ($0 \le F_j \le 1$), in which a gradual measure indicating both the presence and the size of the fault occurring can be incorporated. When the relationship between faults and symptoms is not known, there exist learning techniques in order to establish the correct associations, and to form an explicit knowledge base. Among the techniques in use are statistical or geometrical classification methods (Tou and Gonzales, 1984) and artificial neural networks (Lippmann, 1987; Bishop, 1995; Looney, 1997; Fortuna *et al.*, 2001).

If the relationships between faults and symptoms are partially known, the causal chain:

fault → events → symptoms

can be established. On the basis of this relationships, the heuristic knowledge can be formalized in the form of rules (Frost, 1986; Torasso and Console, 1989) like

IF <condition> THEN <conclusion>

This leads to the so-called fault-tree analysis (FTA). This is a knowledge-based scheme, based on trees, in which symptoms, events, and faults are considered as binary variables, and the rules are established via the application of Boolean equations for parallel–serial connections (Barlow and Proschan, 1975). Recently, approximate reasoning methodologies have received increasing interest to deal with approximate causality knowledge (Isermann and Ulieru, 1993).

The rules obtained in a knowledge-based automatic reasoning system can be chained in two ways. The straighter one is forward chaining. Known facts are matched with the premises and conclusions are derived on the logical consequences imposed by the rules (*modus ponens*). When approximate reasoning is considered, the symptoms are considered as uncertain indicators of faults, with a related degree of truth. This can be taken into account by using confidence functions, fuzzy sets, or probability density functions. On the other hand, backward chaining assumes the conclusions to be known, and searches for all relevant premises (*modus tollens*). This method is particularly useful when symptoms are incomplete or not accurate enough. In practice, forward chaining is usually performed as a first attempt. Then, the user is informed with the possible events and faults that can have occurred. Diagnosis can be further refined by performing a backward chaining that takes into account, as hypotheses, the most plausible events and faults. This procedure can be re-iterated, by interacting with the user, and is usually terminated by the user himself (Freyermuth, 1991).

Another distinction in the automatic reasoning domain has to be made between probabilistic and possibilistic reasoning. The former approach is usually based on Bayesian networks, with associated conditional probabilities for the causality relationships (Pearl, 1988). With this approach, in order to reduce the computational effort, one should consider statistically independent symptoms. On the other hand, possibilistic reasoning is not based on probabilities, but on the concept of degree of truth of logic predicates. This is the basis for the fuzzy logic paradigm. The features or symptoms are represented by linguistic variables, and related to faults via a set of if-then fuzzy rules (Jang, Sun and Mizutani, 1997; Isermann and Ulieru, 1994; Fortuna *et al.*, 2001).

9.5 Trends in Industrial Applications

The survey by Isermann and Ballé (1997) illustrates the trends in fault detection and diagnosis up to 1996, by analysing more than 100 papers submitted to international conferences, dealing with applications on real processes (thus excluding all simulation work). They concluded that in nearly 70% of the work examined, observers and parameter estimation were adopted. Neural networks and parity space, or combined methods were significantly less applied. Moreover, linear models were used much more than nonlinear ones. According to the common trend in every field of system engineering, the number of nonlinear process applications studied through nonlinear models has been clearly increasing over the years, while nonlinear process applications with linearised models have been decreasing. On the other hand, the most common use of soft computing tools, like neural networks and fuzzy logic, up to 1996 was for fault classification and diagnosis (Leonhardt and Ayoubi, 1997), whereas little use of neural networks for residual generation or system modeling was made at that time.

Recent works from the second half of the 1990s dealt with the intrinsic nonlinear nature of systems, introducing nonlinear tools for modeling and fault detection, especially artificial neural networks and fuzzy logic. At present, neural networks are used to build nonlinear models or nonlinear observers, which substitute their linear counterpart in previous approaches. Relevant work has recently been carried out through these approaches (Polycarpou and Vemuri, 1995; Borairi and Wang, 1998; Alessandri and Parisini, 1997; Vemuri, Polycarpou and Diakourtis, 1998; Demetriou and Polycarpou, 1998; Maky and Loparo, 1997; Marcu and Mirea, 1997).

Fuzzy logic is often used for fault classification (Ballé, 1998), implementing the possibilistic reasoning described above. Nevertheless, work has been carried out in which artificial neural networks and fuzzy logic are used as a valuable nonlinear modeling tool (Bucolo *et al.*, 2002; Fortuna *et al.*, 2003; Caponetto, Fortuna and Rizzo, 2003; Esposito, Fortuna and Rizzo, 2004; Maione, Lino and Rizzo, 2005).

9.6 Fault Detection and Diagnosis: A Hierarchical View

Moving to a higher level of abstraction, a hierarchical framework for fault detection and diagnosis systems can be defined. The FDD system can be seen as a hierarchy of machines (as for each level a completed fault detection, diagnosis, and preferably recovering activity is supposed to be implemented), where only contiguous levels can communicate through bidirectional channels.

Three main levels can be identified in a fault detection and diagnosis system:

- Sensor validation: the function of this level is to provide reliable measurements to the upper level, through integration of the available kinds of redundancy. This level should also be able to perform recovery or reconfiguration actions; for example, to exclude faulty measurements, or replace the missing measurement channels with reconstructed variables, obtained by exploiting redundancy and modeling capabilities. At this level, soft sensors are valuable tools to enhance redundancy. Sensor validation will be dealt with in detail below.
- Process fault detection: this level should provide relevant information about the correct behaviour of the process (*i.e.* the plant), or its subsystems. In this level, model-based fault detection and diagnosis can be performed.

- State detection and supervision: this level is often a decision support system (DSS) to operators, with the aim of achieving the best performance within the current operational scenario.

9.7 Sensor Validation and Soft Sensors

Sensors are very important components in any application, as they provide the data needed to implement all the required control, supervision, coordination, optimization, and management activities. Before the 1990s, they were considered as mere signal generators, and it was assumed that measurement errors could be overcome by the feedback loop (Henry and Clarke, 1993). At present, reliability of measurements is considered a fundamental requirement in order to achieve:

- better product quality, plant efficiency and availability;
- enhanced safety or environmental protection;
- better performance of feedback control loops.

The consequences of a sensor fault strictly depend on the application considered: they could range from a reduction in performance or product quality, up to an environmental disaster. It is at present well known that acting on data provided by faulty sensors can lead to much design and optimization effort being wasted. In many applications the only expedient adopted is to replicate sensors, thus achieving physical redundancy. Unfortunately, this is not enough, as it does not ensure measurement quality, reliability, and availability: in fact, it is nearly useless to provide a set of redundant measurements without knowing which sensor to rely on during a specific activity.

The need to apply fault detection and diagnosis strategies to sensors is strongly felt in industrial application. When applied to sensors, fault detection and diagnosis is called sensor validation (SV). Sensor validation becomes challenging when measurements are part of a feedback loop, as the feedback control tends to compensate measurement deviations, thus attempting to hide the sensor fault.

A recent trend in sensor design is to integrate validation activity at the system level. That is to say, a sensor is designed that is able to provide in real time both measurements and extra information about the measurement reliability. This extra information can be exploited in the control loop to perform special control actions in the case of faulty measurement, or at a higher level to schedule maintenance and repair. This kind of sensor design took the name of SElf VAlidating sensor (SEVA) (Henry and Clarke, 1993).

Almost any of the fault detection and diagnosis techniques reported in the literature can be applied to sensor validation:

- Physical Redundancy. Process fault detection cannot generally rely on physical redundancy, because replicating whole parts of the process is very expensive and ineffective. On the contrary, a designer of sensor validation tools can often exploit physical redundancy, as in many industrial applications the installation of redundant sensors is a common fact. It is fair to say that in most industrial applications, the need for redundant sensors is felt and satisfied long before realizing that suitable tools for assessing the reliability of measurements and selecting healthy sensors are absolutely necessary.

- Analytical Redundancy. Sensors can be paralleled by mathematical models which describe the static and dynamic relationship between measurements and sensor inputs, that is to say soft sensors. All the techniques described above for model-based fault detection can be applied to sensor validation. Moreover, on the occurrence of a fault, soft sensors can provide an estimate of missing measurements, at least during a finite time horizon.
- Knowledge Redundancy. As in process fault detection, knowledge redundancy can be exploited to develop qualitative models for sensor validation. Knowledge is usually acquired by interviewing human operators or by analyzing historical fault statistics.
- Measurement Aberration Detection. This technique is based on the analysis of only the output of a sensor, without considering its input–output constituting laws. To validate sensors by means of this technique, it is necessary to assume that the fault-free sensor output has specific properties in the time and/or frequency domain, and that the occurrence of a fault causes changes in these properties. Concerning sensor validation, this approach was first introduced in Yung and Clarke (1989). Also, this technique has a correspondent in the process fault detection domain, which is fault detection with signal models. Many measurable process signals show peculiar features, like oscillations at certain frequencies, biases, and so on. The basic assumption of fault detection with signal models is that the occurrence of a fault causes changes in these features. Therefore, a mathematical model for the signals under analysis is built in order to extract the features from the actual signals and compare them with their nominal value. The related symptoms are therefore analytical.

Effective sensor validation tools are often designed by combining the techniques described above. Very often, the use of a certain form of redundancy gives better results in the detection of some faults, whereas another form might be exploited more effectively in order to detect other kinds of faults.

One of the approaches to sensor validation presented in this chapter, is to merge different sensor validation techniques in a knowledge-based framework. Within this approach, the partial results achieved by single validation techniques devoted to the effective detection of specific faults, are evaluated by a fuzzy inference system (FIS) able to perform a judgement about the measurement quality from a global point of view.

9.8 Hybrid Approaches to Industrial Fault Detection, Diagnosis and Sensor Validation

Although many efforts have been made to set a general methodology for fault detection and diagnosis, it has been recognized that when attention is switched to industrial problems, the related solutions become more and more customized to the specific application. After a thorough analysis on a large number of industrial cases, some common issues can be highlighted:

- most of the phenomena are nonlinear, and the nonlinearities are often not known, or difficult to model;
- physical modeling is hard; very often causal relationships are (partially) known, but exact physical laws governing the phenomena are too complicated for an effective model to be built;
- redundancies are "partial". None of the kinds of redundancy explained previously is really fulfilled. For example, in a chemical plant there can be several sensors, placed at different locations of a circuit, able to measure the concentration of a reagent. In this way, physical redundancy is not strictly speaking achieved, but the measurements are closely related to each other. Moreover, the relation between these measurements can only be set in a qualitative fashion, as physical details about the relation might be poorly known or too complicated. This is a case in which partial physical redundancy and knowledge redundancy coexist. Soft sensors can help in exploiting this kind of redundancy.
- Industries often require diagnostic systems to be installed in a non-invasive fashion, *i.e.* they must exploit the existing resources without any modification. For example, diagnostic tools often have to be designed without installing new sensors, nor changing the sampling rate of measurements or the data format in the measurement database. Moreover, it is often impossible to reconstruct specific scenarios for design or testing purposes, or to simulate faults, *etc.* In fact, the only resource the designer of fault detection and diagnosis tools can access is the measurement data stored in the industry database.
- Measurement databases are often the only resource available, but they are not accompanied by useful additional information, like a log of the occurrence of faults (size, time, location and type of the fault). This kind of information, together with heuristic knowledge, is usually kept in the operator's mind, or in personal logbooks. Consequently, interviewing operators and experts to acquire knowledge redundancy is a key point for an effective design of diagnostic tools.

It is fair to say that, in industrial applications, effective design of a fault detection and diagnosis system must be driven both by experimental data and human experience. We have already stated in this chapter that fault detection and diagnosis can be developed in a knowledge-based framework. The majority of early work (in the 1970s and 1980s) in the field of knowledge-based diagnosis revolved around a manual construction of the knowledge base (Milne, 1987). An underlying assumption of an application-oriented approach is its ability to assemble the knowledge required for a particular task in a cost-effective manner. Although there is a great variety of tasks for which extensive knowledge engineering is justifiable, this is only one class of problems, and is too small to take advantage of the full capabilities of artificial intelligence, and also to build a reasonable quantity of economically viable systems that prove beneficial to industry. Unfortunately, there is a large class of problems for which solutions seem achievable, but the extensive manual knowledge engineering approach is unjustifiably expensive.

These considerations lead to the development of hybrid approaches, like those presented below in this chapter. Hybrid approaches try to conjugate the benefits of knowledge engineering with those of approximate reasoning, nonlinear black-box modeling, and optimization tools. As previously stated, soft computing techniques seem particularly suitable for this purpose. For example, thanks to the ability of neural or neuro-fuzzy networks to interpolate nonlinear functions, the expert (partial) knowledge can be refined by exploiting the presence of stored experimental data. In this way, experimental data can make up for the lack of knowledge about the system behavior.

As an example, the first application presented in what follows is based on a hybrid approach, which exploits the advantages both of signal models and system models for fault detection. Instead of designing a model of the system under analysis, *i.e.* a mathematical tool able to predict the trend of the system output under nominal operational conditions, it has proved possible to exploit human knowledge and experimental data in order to predict directly the nominal signal features on the basis of experimental measurements. Predicted symptoms can then be compared to actual symptoms by means of a diagnosis tool to generate the fault vector.

Within hybrid approaches, the algorithms designed are often based on soft computing tools. Consequently, they are able to incorporate human knowledge (in terms of fuzzy if-then rules) and the ability to learn from data (through neural or neuro-fuzzy tools).

In conclusion, the hybrid approach to fault detection and diagnosis proposed in this chapter can be developed through the following steps:

- **conducting interviews with experts,** in order to acquire the knowledge about the involved phenomena, the known causalities relationships, the faults, their known causes, and all the heuristic knowledge available;
- **defining the symptoms** and the algorithms to derive them from the available measurements;
- **designing suitable detection tools,** able to exploit the different kinds of (partial) redundancy present in the system, the available data and the heuristic knowledge;
- **comparing** the expected symptoms and the actual ones, and evaluating the difference by means of either fixed or adaptive thresholds, or via a **diagnosis tool** (which can also be tuned on the basis of existing diagnosis data stored in the database).

It is worth noting that hybrid approaches do not exclude the presence of traditional fault detection and diagnosis techniques, which can be integrated into the design of a more complex tool. As stated in the previous section concerning sensor validation, also in the case of fault detection and diagnosis the partial results obtained by single fault detection tools can be evaluated and merged by a FIS, in order to provide a diagnosis based on the analysis of the results of individual, specific techniques.

In what follows, two applications developed by means of hybrid approaches will be shown. In particular, the design of two sensor validation tools for nuclear fusion applications will be described. The first is designed to validate a set of measurements of mechanical stress, performed by strain gauges, of the vacuum vessel where nuclear fusion takes place. The second is designed to validate the measurement of plasma density during fusion reactions, performed by an interferometer. The tools developed are currently installed at JET Joint Undertaking (Culham, UK), and at FTU Frascati Tokamak Upgrade (ENEA Frascati, Italy), respectively.

9.9 Validation of Mechanical Stress Measurements in the JET TOKAMAK

The Joint European Torus (JET) is the world's largest fusion facility. It is a project in the coordinated fusion program of the European Atomic Energy Community (EURATOM), whose long-term objective is the creation of a prototype of a fusion reactor, complying with the ever more challenging safety and environmental care requirements. It is now managed by EFDA (European Fusion Development Agreement) and hosts scientists from all over the world, to conduct experimental campaigns in the nuclear fusion domain. Many disciplines are involved in the achievement of the fusion reaction. Consequently, JET, like many other physics laboratories, is also a great test bed for most of the science and engineering fields (*e.g.* robotics, mechanics, electromagnetic fields, power supplies, computer science, algorithms, computer networks, optimization, automatic control, instrumentation and measurements, *etc.*)

At JET, sensor validation is felt to be a primary task to be accomplished, for several reasons, briefly listed below:

- the JET machine has worked for a long time above its nominal design limits, and a further extension of the toroidal field (the field necessary for plasma confinement) to 4 T was carried out in 1999. Thus, the need for reliable diagnostic systems is more and more important due to the high energy involved;
- due to the experimental nature of the JET machine, a remarkable number of sensors has been installed, and is currently increasing, in order to evaluate the machine's performance, improve the availability of measurements (during deuterium–tritium operations access to the operational area to repair a sensor is not allowed), and study nonaxes-symmetric machine behavior;
- the considerable reduction in staff requires a new approach to fault analysis, in which automatic tools, which take human experience and all kinds of redundancy into account, can help operators to monitor thousands of variables;
- testing new techniques in fault detection and sensor validation is very important for use in other plants with a huge number of variables, like the ITER tokamak, other physics laboratories, or industry.

The application of automatic sensor validation techniques is rather new in the field of nuclear fusion and experimental physics, due to the complexity of the phenomena involved and the almost total absence of accurate models of the plant or parts of it. So far, almost all fault detection activity has been carried out manually by experts, with the help of simple signal processing operations, threshold-based alarms, and the experience acquired by plant operators.

The aim of the application presented in this section is to build an automatic sensor validation system for a sensor set which measures the vertical mechanical stresses on the supports of the JET tokamak vacuum vessel. Experts are interested in the reliability of these measurements during specific time intervals, corresponding to the occurrence of particular physics phenomena called disruptions. During the usual operational life of the tokamak, in fact, mechanical stresses are relatively weak and do not need to be monitored. During disruptions, on the contrary, fast dynamic vertical displacement events (VDE) occur, causing a great swing to and mechanical oscillations of the vacuum vessel, which must be monitored to ensure the mechanical integrity of the machine. One of the actions related to the monitoring of the mechanical stresses is to suspend the experimental campaign when more than a fixed number of VDEs exceeding a certain stress threshold occur in any day. The reliability of the measurements is therefore very important to avoid both unnecessary suspension of experiments and dangerous experiments carried out beyond operational limits. In what follows, the design of the sensor validation tool is described. After a brief description of the measurement system, a classical approach based on physical redundancy is presented. Subsequently, a hybrid sensor validation system designed according to the guidelines described in Section 9.8 is presented and, finally, validation results are evaluated and compared.

The measurement system under analysis is a set of 32 strain gauges located along the vertical port restraints of the vacuum vessel. Measurements taken by these sensors are very significant, as they properly reflect the mechanical behaviour of the machine during disruptions. Vertical forces measured through these strain gauges are subsequently used to compute the overall stress of the vessel at disruption. It is obvious that the inclusion of faulty measurements into the computation algorithm lead to misunderstandings of the physical phenomena that occur.

The details concerning the process and the measuring system are given in the Appendix.

9.9.1 Heuristic Knowledge

Under the occurrence of a plasma disruption, plasma confinement is suddenly lost and the plasma current falls to zero in a very short time. Consequently, high mechanical and thermal stresses are produced in the machine structure (Buzio *et al.*, 1996). Disruption-induced loads are characterized by radial and vertical components of several millions of newtons, with typical timescales ranging from 20 to 50 ms. Since the installation of additional restraining rings (1989), the vessel has become more rigid with respect to radial axisymmetric forces and the most important mechanical loads are now due essentially to vertical loads. Because of the particular configuration of the supports, vertical forces exert a torque around a rotation centre, generating a vessel axisymmetric motion (called rocking motion by the experts), with a frequency of oscillation at about 14 Hz. Experts summarize the force history into the following stages (Buzio *et al.*, 1996):

1. initial steady phase;
2. upward force;
3. large downward swing;
4. slow decay at 14 Hz.

The downward swing occurring in phase 3 is the most important with regard to the fatigue life of the machine. Experts estimate this swing by computing the so called F-number (JET, 1998), a nonlinear function of seven currents, which are part of the settings of the experiment. More details about these currents are given in the Appendix.

The vessel in which the reaction takes place has a toroidal shape, and is divided into eight octants. Strain gauges are replicated in each octant (eight sensors per octant, four are installed in the upper restraints and four in the lower restraints), and experiments are generally symmetrical along the toroidal direction. Therefore, the measurements performed should exhibit this symmetry. Even though measurements along the vessel are strongly related by toroidal symmetry relationships, this cannot be strictly considered a case of physical redundancy, where multiple sensors are installed in order to perform exactly the same measurement. This has already been referred to in the previous section as a case of partial physical redundancy.

In the next section, a simple algorithm to exploit this kind of redundancy is described.

9.9.2 Exploiting Partial Physical Redundancy

In nominal operational conditions, the 32 lower strain gauges, which are in principle related by toroidal symmetry, should perform very similar measurements at the same time, apart from the initial offset, which depends on sensor calibration. The following approach allows us to check whether a sensor is performing a measurement which does not respect the symmetry criterion. This is a model-free approach, in which the reference behavior is not generated by a model.

The algorithm is based on the computation of an average behavior (Dorr *et al.,* 1997) of the sensor set. Subsequently, the difference between each sensor output and the average behavior is evaluated. Even if this is a model-free approach, this difference accomplishes the same function as a residual. Subsequently, if the residual is too big according to some criterion, a fault condition for the corresponding sensor is raised.

The algorithms can be formalized as follows:

- consider the measurement performed by the set of n sensors

  $x_i(t)$, with i=1,2, …, n

- subtract from each measurement its initial offset

$$x_{d_i}(t) = x_i(t) - x_i(0) \tag{9.34}$$

- compute the residuals

$$r_i(t) = x_{d_i}(t) - \frac{1}{n}\sum_{j=1}^{n} x_{d_j}(t) \tag{9.35}$$

Once the residuals 9.35 are generated, they must be checked according to a fault detection criterion.

The first criterion which can be adopted is to fix a threshold K and time window T. An alarm is then raised if $|r_i(t)|>K$ during the whole time window T. The time window T is set to add robustness to the detection, thus avoiding a false alarm from being raised if the threshold is exceeded for a very short time interval. Obviously, the choice of threshold and time window is based on a trade-off between accuracy and robustness.

As previously stated, strict physical redundancy of sensors is not available, and the algorithm is based on the assumption that the sensors should be measuring the same quantity. It can happen that, in certain experiments, toroidal symmetry can be less evident. In this case, with a fixed threshold, most of the sensors may be declared faulty due to the lack of symmetry of the specific experiment, whereas they are performing a correct measurement. Therefore, an adaptive threshold may be more suitable. A lower threshold is needed when sensor outputs are similar, whereas a higher one must be fixed in order to tolerate cases in which symmetry is missing. With this in view, let us consider

$$\sigma(t) = \sqrt{\frac{1}{n}\sum_{i=1}^{n}\left(r_i(t) - \bar{r}(t)\right)^2} \tag{9.36}$$

where

$$\bar{r}(t) = \frac{1}{n}\sum_{i=1}^{n} r_i(t) \tag{9.37}$$

Then, a fault condition on the ith sensor is raised if $|r_i(t)|>k\sigma(t)$ (with $k>0$) during a fixed time window T. In our application, satisfactory results are obtained with $2<k<3$ and a time window between 10 and 50 samples of the sensor output. Fixing a time window can be avoided if integral functions of residuals and adaptive threshold are compared, that is checking whether

$$\int_0^t |r_i(\tau)|d\tau > k\int_0^t \sigma(\tau)d\tau \tag{9.38}$$

This can be a suitable approach to avoid fixing a time window and to cope with different kinds of faults, like big spikes over short time intervals, or constant, small biases over long time intervals. Moreover, the detection performance may be improved by introducing a forgetting factor, *i.e.* a low-pass filter with a suitable time constant instead of a pure integrator.

This approach is simple to implement. Nevertheless, it is suitable for coarse fault detection, and certain kinds of fault cannot be easily detected, unless the sensitivity of the algorithm is increased, thus leading to the frequent raising of false alarms. Moreover, this algorithm can perform fault detection, but is unable to perform fault classification or diagnosis. In the next section, a hybrid sensor validation strategy for the same set of sensors will be described (Rizzo and Xibilia, 2002; Fortuna *et al.*, 1999), based on the hybrid methodology described in Section 9.8.

9.9.3 A Hybrid Approach to Fault Detection and Classification of Mechanical Stresses

The design of the sensor validation tool follows the steps described in Section 9.8. A functional scheme for the tool is illustrated in Figure 9.5. The algorithm hybridizes the signal model with model-based techniques. After having acquired knowledge about possible sensor faults, adequate signal features for sensor fault detection are defined. Actual signal features can easily be computed from measurements with standard signal processing operations. On the other hand, unlike fault detection with signal models, nominal signal features cannot be fixed, as JET is an experimental physics plant and experiments can be very different from each other. Consequently, the fault-free behavior of sensors can be very changeable. Therefore, nominal features depend on the specific experiment. To overcome this problem, nominal features are predicted with a neural model trained on experimental data, thus obtaining a nominal feature model, which can be fed either by the sensor inputs or, as in this application, by other quantities related with the sensor response. Subsequently, actual and predicted nominal features are compared to generate analytical symptoms. Finally, symptoms are evaluated by a fuzzy inference system able to perform the validation.

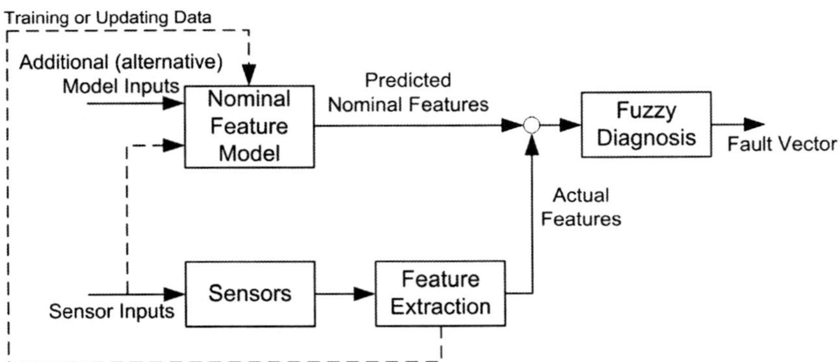

Figure 9.5. Conceptual scheme for the sensor validation of mechanical stresses at JET

9.9.3.1 Knowledge Acquisition

As already stated in Section 9.9.1, plasma disruption produces a vertical displacement event (VDE) of the vacuum vessel, followed by a so-called rocking motion at a frequency of about 14 Hz. The characteristic downward swing is a function of seven currents, which are part of the setting of the experiments. The set of 32 strain gauges is installed to monitor this mechanical stress when a disruption occurs, and sensor faults must be correctly detected and classified, in order not to include faulty measurements in the computation of the overall stress of the mechanical structure.

After a thorough investigation, four classes of faults for the installed strain gauges were identified:

- **gain fault,** in which the faulty signal denotes the same shape as the non-faulty ones, but it is magnified or scaled by a multiplicative factor; this is caused by a sudden change of gain in the acquisition system;
- **bias fault,** in which the faulty signal has the same shape as the non-faulty ones, but its trend is broken in several parts separated by steps; this is caused by sudden introduction of constant biases in the measurement;
- **spike fault,** in which the faulty signal presents large and occasional spikes; this can be caused by bad adherence of the strain gauge to its support;
- **noise fault,** in which high frequency noise is superimposed on the measurement. This noise is occasionally caused by certain radiofrequency amplifiers installed in the machine (fast radial field amplifier, FRFA).

It has been found that noise and spike faults never occur simultaneously in the same measurement.

9.9.3.2 Symptom Definition

As stated before, the approach adopted is a hybridization of model-based fault detection with signal models. In particular, the model developed allows the nominal value of the signal features to be predicted on the basis of the operational condition, in order to compare them with the actual feature values. Analyzing the waveforms of typical sensors responses, four features were reckoned to be sufficient to isolate and classify the faults described above. Denoting with $x_i(k)$ the sensor time-discrete output of the *i*th sensor and with k the discrete time index, they can be listed as follows:

- **Maximum peak** of the oscillations which, together with the next symptom, is a suitable indicator of gain and bias faults;

$$M_i = \max_k x_i(k) \tag{9.39}$$

- **Average value** of the sensor response (where N_s is the total number of samples of the sensor output during an experiment);

$$A_i = \frac{1}{N_s} \sum_{k=1}^{N_s} x_i(k) \tag{9.40}$$

- **Maximum of the absolute value of the prime difference function**, which is a suitable indicator for the detection of spike faults;

$$P_i = \max_k \left| x_i(k) - x_i(k-1) \right| \tag{9.41}$$

- **Sum of the magnitude of FFT samples between 100 and 130 Hz,** which is a suitable indicator of the detection of noise faults. Since $X_i(f)$=FFT$\{x_i(k)\}$, it is

$$N_i = \sum_{f=100Hz}^{300Hz} \left| X_i(f) \right| \tag{9.42}$$

9.9.3.3 Design of the Detection Tool: Nominal Features Model

The aim of this subtask is to build a hybrid model able to predict the nominal features of the sensor response, on the basis of the specific operational condition. With this aim in view, a neural model is designed to establish directly the correspondence between the causes of the mechanical stress and the nominal features expressed in Equations 9.39 to 9.42. We have already mentioned that the main mechanical event after the occurrence of a disruption, which is the large downward swing, is caused by the value of seven currents, which are part of the setting of the experiment. Moreover, the swing is determined by the instantaneous value of the seven currents at the disruption time (for practical reasons explained in the Appendix, the instantaneous value is actually taken 200 ms before the occurrence of the disruption). Therefore, as the mechanical behavior of the system is determined by these currents, we conjectured that it is possible to find a correspondence between the instantaneous value of the currents at the disruption time and the features 9.39–9.42 in fault-free conditions.

To design the hybrid model, an MLP was used (Fortuna *et al.*, 2001). The MLP adopted has therefore seven inputs and four outputs. By trial and error, the size of the hidden layer was fixed at two neurons. As our aim is to build a feature model for a healthy sensor, the network was only trained on healthy data. With this purpose in view, the manual validation data log was browsed, and calculation of the correct values of the features was performed on the basis of manually validated strain gauge measurements. The MLP was trained on 60 experiments, covering most operational scenarios. A very large checking set, comprising 200 experiments, was selected to validate the model for a wide range of situations. Figure 9.6 shows the distribution of errors made by the network on the four features, for the whole set of 260 experiments.

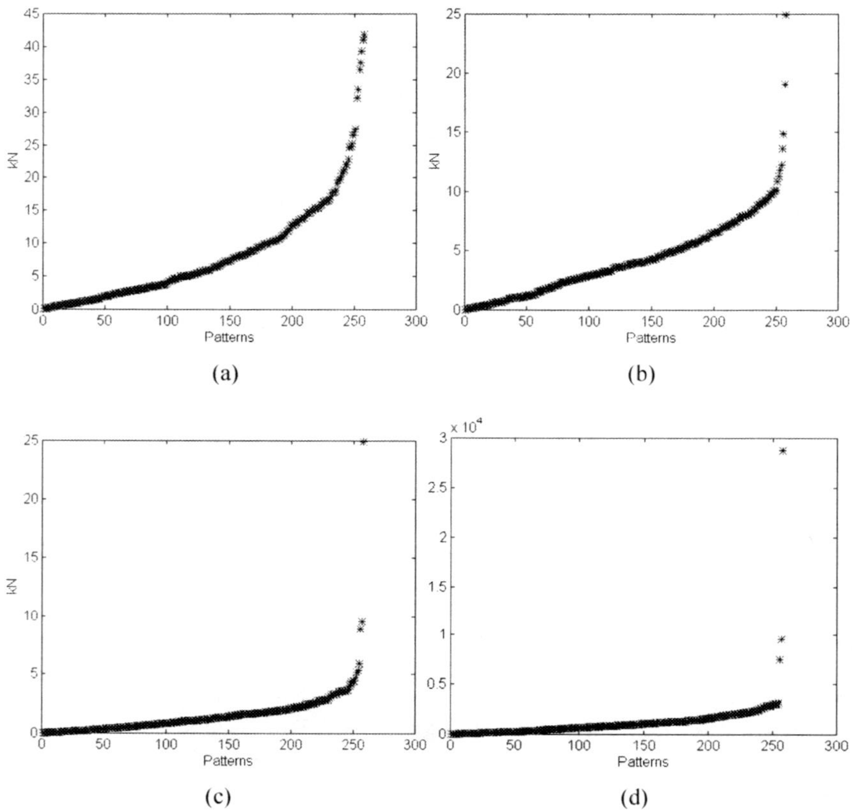

Figure 9.6. Error distribution of the nominal features model, for the features expressed in Equations 9.39–9.42, respectively

As can be seen, this distribution is quite uniform, except for a few experiments, belonging to the checking data set, in which the error is extremely high. After a thorough analysis of the patterns on which the error made by the model is high, it was recognized that this behavior is caused by mistakes made by the automatic system for detection of the disruption time. As a consequence, the model input was not evaluated at the correct time for these experiments. Although, in principle, these experiments should not have been taken into account in the checking set, in the real application this fact can occur, even though with a low probability. In these cases the sensors may be detected as faulty (false alarm).

The difficulty of detecting the correct disruption time with an extremely high accuracy depends on the nature of the phenomena involved. In fact, a disruption is detected on the basis of observation of some precursors. Sometimes the precursors observed do not lead to an actual disruption, or do not reveal the main (strongest) disruption. As the current quenches during a disruption are very steep, a small error made in evaluation of the disruption time involves a large error in the actual current values, which cannot be related to the chosen features, leading to model misbehavior. In any case, the patterns denoting an unacceptable modeling error constitute about 5% of the whole pattern set. In validating the model, the maximum error figures obtained in the remainder of the measurements (thus constituting 95% of the total) are given in Table 9.1.

Table 9.1. Maximum estimate errors made by the neural symptom model and range of measurement

Feature	Max. Err.	Range
M_i	16 kN	400 kN
A_i	8 kN	400 kN
P_i	3 kN	400 kN
N_i	1000	10000

The figures given in Table 9.1 confirm the satisfactory performance of the model, considering that strain gauges have a measurement scale ranging from 0 kN to 400 kN. The maximum error of 16 kN is therefore less than 5% of the measurement range. Concerning the last parameter, in Table 9.1 the error made is quite small, given that it has been checked that the indicator N_i for a noisy signal differs from that related to a non-noisy one by at least 5000.

Once the features M_i, A_i, P_i, N_i are predicted, they are compared with the corresponding features computed from the actual sensor outputs. The differences between predicted and actual features are the analytical symptoms which have to be analyzed for fault diagnosis, as will be shown in what follows.

9.9.3.4 Fuzzy Fault Diagnosis

The symptoms generated can be analyzed using several methodologies, like one of those mentioned in Section 9.4. As Figure 9.5 shows, the approach presented in this chapter is modular, with separated detection and diagnosis functional blocks. In this application, a fuzzy inference system is adopted, because a linguistic model for fault diagnosis is available. The linguistic model is based on both the analysis of the symptoms generated and investigation of the expert knowledge. As mentioned previously, four main faults can be isolated. For the same sensor, they can occur either alone or concurrently, on the basis of the various combinations of the symptoms computed by the neural model.

To perform the fuzzy diagnosis, a Takagi–Sugeno fuzzy system was chosen (Jang, Sun and Mizutani, 1997). Let **PKerror, AVerror, SPerror,** and **NSerror** be the difference between the actual and estimated features expressed in Equations 9.39 to 9.42. These differences are the inputs of the fuzzy inference system. Each input has three associated membership functions, called **pos, zero,** and **neg,** defining the fuzzy concepts of positive, negative, or null error. The shapes and overlaps of the membership functions are fixed by trial and error, considering heuristics, performance, and sensor and model accuracy.

The four faults described in the previous section can be classified through three indicators, **Amplification, Bias,** and **Disturbances**, which are the outputs of the fuzzy system. Each of the outputs can assume three crisp values:

- **Amplification and Bias:** high (1), low (−1), ok (0);
- **Disturbances:** noise (−1), spikes (1), ok (0).

The indicator **Disturbances** accounts for the diagnosis of both spike and noise faults because, as mentioned previously, these faults do not occur simultaneously in the same measurement.

The following set of rules was designed:

```
1. if PKerror is pos and AVerror is zero then Ampli
   is hi and Bias is ok
2. if PKerror is neg and AVerror is zero then Ampli
   is lo and Bias is ok
3. if PKerror is not neg and AVerror is pos then
   Ampli is ok and Bias is hi
4. if PKerror is not pos and AVerror is neg then
   Ampli is ok and Bias is lo
5. if PKerror is zero and AVerror is zero then Ampli
   is ok and Bias is ok
6. if SPerror is not pos and NSerror is notpos then
   Disturb is ok
7. if SPerror is zero and NSerror is pos then Disturb
   is noise
8. if SPerror is pos and NSerror is pos then Disturb
   is spikes
```

Rules 1 to 4 allow the presence of amplification and bias faults to be revealed, analyzing the error on both the maximum peak and the temporal average. As an example, Rule 1 says that if the error on the peak is positive and the error on the average is zero, then the signal is magnified but not biased, *i.e.* this is the case of a gain error. Rule 5 describes the condition of a sensor not affected by faults. Finally, Rules 6 to 8 deal with the diagnosis of spikes or noise faults.

9.9.3.5 Performance Assessment

The sensor validation tool was implemented in Matlab® on a UNIX platform. A short Fortran routine was needed to retrieve signals from the JET database.

In order to compare the performance of the hybrid approach and that based on partial physical redundancy, explained in Section 9.9.2, both algorithms were run on a new set of 96 experiments, where manual validation performed by experts found that four strain gauges were actually affected by different kinds of faults.

A comparison between the two methods is shown in Table 9.2, where the average number of faulty measurements over the total number of measurements (number of sensors multiplied by number of experiments considered) is given. This can also be considered as the average number of faulty sensors per experiment (denoted as AFSE in Table 9.2).

Table 9.2. Comparison between the two strategies: based on physical redundancy (Phys. Red.) and hybrid

Method	Total (Number of sensors × number of experiments)	Number of faults	AFSE
Phys. Red.	3072	51	0.5
Hybrid	3072	375	3.90

From Table 9.2, an AFSE of 3.9 faulty sensors per pulse obtained through the hybrid approach is realistic. In fact, as stated above, in the period considered, four sensors were faulty. On the other hand, the figure of 0.5 provided by the classical method, referring to the same operational period, is incorrect. Moreover, significant improvements in the traditional method were not achieved with further tuning of the thresholds and time window.

The AFSE is a suitable performance index with regard to the quantitative aspects of the detection task. An in-depth analysis through the single experiments revealed that the sensors detected as faulty by the hybrid approach correspond to those really affected by faults. The diagnosis task was manually validated by examining the sensors declared faulty, providing good performance in all cases.

In conclusion, the results obtained by the hybrid approach were considered comparable to those obtained by thorough manual validation performed by plant experts.

9.10 Validation of Plasma Density Measurement at ENEA-FTU

In this section, we deal with the design of a sensor validation tool for the plasma density measurement of the FTU (Frascati Tokamak Upgrade) Tokamak, located in Frascati (Rome), Italy, at the ENEA Laboratories. While mechanical stress monitoring treated in the previous section is essential for machine safety, correct sensor validation of plasma density measurement is essential for the evaluation of the physics results achieved by the nuclear fusion experiment (Buceti *et al.*, 2001a; Buceti *et al.*, 2001b; Buceti *et al.*, 2002). Plasma density measurements are performed by a deuterium–carbon–nitrogen (DCN) interferometer, an instrument able to perform the measurement of plasma electron line density. It is a five-channel instrument, in which each channel performs a measurement along a different chord of the plasma column (which has a circular section). More details about interferometry and plasma density measurement are given in the Appendix.

As seen below, the validation system designed fits well into the hybrid development framework described in Section 9.8, as model-based validation, signal model validation, knowledge-based redundancy, physical redundancy, are concurrently exploited. The main idea underlying the design of the validation tool is to perform single validation operations by using different strategies, each specialized in the analysis of specific aspects of the measurement. Subsequently, the results from single validations are merged by a fuzzy inference system to draw global conclusions about the quality of measurement. One of the techniques adopted to perform fault detection is based on the development of a soft sensor, used to implement a model-based fault detection strategy. The soft sensor in this application is implemented by means of an MLP trained on experimental data. The design and development of the tool is described below, through the development steps outlined in Section 9.8.

9.10.1 Knowledge Acquisition

During a fusion experiment, plasma density is evaluated by integrating the line densities measured by the DCN interferometer along five parallel chords. After a thorough analysis of the physical phenomena involved, and interviews with experts, the following considerations can be listed:

1. plasma line density is mainly related with the plasma current and with the amount of gas injected into the vacuum vessel;
2. the plasma column is denser near its centre; this implies that line density measured along lines which are far from the plasma centre is smaller than line density measured along lines which are near the plasma centre;
3. line density measurements have a similar trend along all lines (and also similar to the plasma current trend);
4. line density measurements should not be affected by important discontinuities, like steps or spikes; steps can be frequent because they are caused by a well-known sensor fault, typical of interferometers, called fringe skip (see Appendix).

It can be noticed that the availability of five line measurements along five different chords of the same plasma column constitutes a case in which partial physical redundancy exists. Moreover, a cause–effect relationship between plasma density and its main causes can be established. Therefore, this knowledge redundancy can lead to the development of a model, thus achieving analytical redundancy. Finally, some additional knowledge redundancy is available, like considerations of the measurement trends or the presence of discontinuities.

A block scheme of the validation tool is illustrated in Figure 9.7

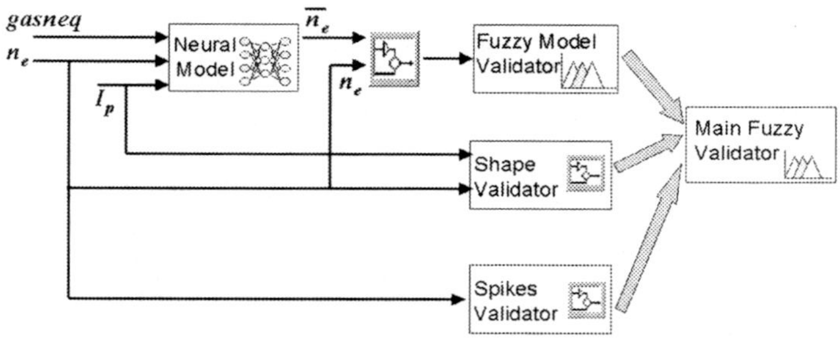

Figure 9.7. Block scheme of the sensor validation system for plasma density measurements

9.10.2 Symptom Definition

The functional blocks of the scheme are illustrated in Figure 9.7. The block called Main Fuzzy Validator consists of a fuzzy inference system able to evaluate symptoms generated by the blocks placed at its inputs. It has the same function as the fuzzy inference system described in Section 9.9.3.4. Three different symptoms are generated and evaluated:

- The first symptom refers to the first point in the list of acquired knowledge, which establishes a causal relationship between plasma line density and other variables. The symptom is generated by considering the residual of a model for the plasma line density. The model developed is nonlinear, black-box, based on a MLP neural network, with a NARX structure. Two different analytical symptoms are generated (which are two features of the model residual), and subsequently analyzed by another fuzzy inference system, called Fuzzy Model Validator in Figure 9.7, which performs a first analysis and generates a symptom which will be the input of the Main Fuzzy Validator.
- The second analytical symptom refers to the third point of acquired knowledge, which establishes a relationship between the temporal trends of plasma current and plasma density. It is denoted X_c and is computed as

$$X_c = \max(xcorr(I_p, d)) \tag{9.43}$$

  where I_p is the plasma current, d is the plasma line density and *xcorr* is the normalized cross-correlation function. This indicator assumes high values as long as the density measurement and plasma current have similar temporal trends.
- The third symptom takes into account the fourth point of the knowledge basis, and is designed to detect the occurrence of spikes or discontinuities on the sensor output. In order to compute this indicator, a high-pass filter is applied to the sensor output, and the maximum of the resulting signal is considered to be an analytical symptom. This value is high as long as the measurement is affected by discontinuities.

In the following section, the neural model, the fuzzy model validator, and the main fuzzy validator are described in greater depth.

9.10.3 Design of the Detection Tool: Soft Sensor and Fuzzy Model Validator

As stated in Section 9.10.1, plasma line density is mainly related with the plasma current and the amount of gas introduced into the tokamak. A physical model cannot be designed as the physical relations between variables are not available, therefore a NARX model is developed in order to obtain a soft sensor for the plasma density. The model structure is:

$$d(k+1) = f\big(I_p(k+1), I_p(k), gas(k+1), gas(k), d(k), d(k-1)\big) \tag{9.44}$$

where k is the discrete time index, $d(k)$ is the plasma line density, $I_p(k)$ is the plasma current, and $gas(k)$ is the total amount of gas introduced up to time k.

The nonlinear model was identified using a MLP neural network with six inputs and one output. By trial and error, the number of neurons in the hidden layer was set at nine. The data set used to train the model was built using data selected from the experimental database of FTU. To build the nominal model, only data manually validated by experts were considered. Moreover, they were selected in order to cover as many operational scenarios as possible.

The model performance was subsequently tested by a new set of experimental data. This consists of 177 experiments in which the sensors are not affected by faults, and 227 experiments in which sensors are faulty. The average modeling error, computed on the test data set and normalized in the range [0,1] is 5.4×10^{-5} on the fault-free cases, and 2.1×10^{-2} on the faulty data set. The great difference between the average error made by the model on the fault-free data and that made on the faulty data indicates that the model is able to distinguish between faulty and fault-free data.

It may happen that the residual related to a fault-free signal exceeds a fixed threshold for a short time, or by a small amount. Moreover, it is hard to fix a threshold with a good trade-off between accuracy and robustness. To reduce the possibility of generating false alarms and provide a more reliable symptom for the final validation stage, two features of the residual are analyzed by the fuzzy model validator. They are:

- the average value of the residual over the whole experiment (called *avgerr* in what follows);
- since σ is the standard deviation of the model error on the whole set of fault-free data, the second parameter is the maximum number of samples in succession out of a bound of $\pm 2.5\sigma$ (called *outbound* in the following).

The Fuzzy Model Validator takes as inputs the two variables *avgerr* and *outbound.* Five linguistic values have been defined for the variable *avgerr: neghigh, negative, zero, positive, poshigh;* and three for the variable *outbound*: *small, medium, high.*

The following set of rules has been designed. The output fuzzy variable *model* is the symptom which is fed as input to the Main Fuzzy Validator.

```
1. if Avgerr is Zero and Outbound is Small then Model
   is Ok
2. if (Avgerr is Positive or Avgerr is Negative) and
   Outbound is Small then Model is Warning
3. if (Avgerr is Poshigh) or (Avgerr is Neghigh) then
   Model is Faulty
```

```
4. if (Avgerr is Negative) or (Avgerr is Zero) or
   (Avgerr is Positive) and Outbound is High then
   Model is Faulty
5. if (Avgerr is Negative) or (Avgerr is Zero) or
   (Avgerr is Positive) and Outbound is Medium then
   Model is Warning
```

It can be seen from the set of rules that the output variable *model* can assume three fuzzy values: *Ok, Warning, Fault.* The *Warning* value has been introduced to highlight cases in which computation of the symptom leads to an uncertain result.

9.10.4 The Main Fuzzy Validator

The Main Fuzzy Validator block was designed to take into account the three symptoms previously described, and to draw conclusions about the status of the sensor on the basis of the symptoms fed in at its inputs. It consists of a fuzzy inference system with three inputs and one output. The inputs are the three symptoms generated by the Fuzzy Model Validator, the check on the cross-correlation between plasma line density and plasma current, and the check for the presence of discontinuities. The three symptoms are called *Model*, X_c, and *Spikes*, respectively. Each of them has a set of associated linguistic (fuzzy) values, as follows:

1. *Model:* Ok, Faulty, Warning;
2. X_c: Low, Medium, High;
3. *Spikes:* Low, Medium, High, Very High.

The output variable of the fuzzy inference system is called *Sensor,* and can assume three values: *Ok, Warning, Fault.* The variable *Sensor* indicates the final validation result. On the basis of the experts' knowledge, according to a linguistic description of the manual validation process, the Main Fuzzy Validator rules have been designed as follows.

```
1. if Xc is High and Model is Ok and Spikes is Low
   then Sensor is Ok
2. if Xc is Not(High) and Model is Faulty and Spikes
   is Not(Low) then Sensor is Faulty
3. if Xc is Low and Model is Ok then Sensor is
   Warning
4. if Xc is High and Model is Warning and Spikes is
   Low then Sensor is Ok
5. if Xc is Medium and Model is Warning then Sensor
   is Warning
6. if Model is Warning and Spikes is Not(Low) then
   Sensor is Faulty
7. if Xc is High and Model is Faulty and Spikes is
   Low then Sensor is Faulty
```

```
8.  if Model is Ok and Spikes is Not(Low) then Sensor
    is Warning
9.  if Xc is Low and Model is Ok then Sensor is
    Faulty
10. if Model is Ok and Spikes is VeryHigh then Sensor
    is Faulty
11. if Xc is Medium and Model is Ok then Sensor
    Warning
12. if Model is Ok and Spikes is High then Sensor is
    Warning
13. if Xc is Small and Model is Warning then Sensor
    is Faulty
14. if Xc is Medium and Model is Faulty and Spikes is
    Not(Low) then Sensor is Faulty
15. if Xc is High and Model is Warning and Spikes is
    Medium then Sensor is Warning
```

9.10.5 Performance Assessment

The performance of the validation system has been assessed on a set of 300 experiments, containing either faulty or non-faulty measurements, manually validated by experts. These experiments were not used in the design phase. Validation results are shown in Table 9.3. The second and fifth columns (*i.e.* the columns labeled Ok→Ok and Fault→Fault) report the success rate of the tool in sensor validation and fault detection, respectively. The second column reports the percentage of successful validation, that is the number of measurements detected by the validation tool as not affected by fault, with respect to the total number of fault-free measurements. The fifth column reports the percentage of successful fault detections, that is the number of measurements detected by the validation tool as faulty, with respect to the total number of faulty measurements. The third and fourth columns (Ok→Fault and Ok→Warn) report the percentage of failures in the validation process, which is the percentage of misclassification of fault-free measurements, classified as being in a faulty or warning state, respectively, by the automatic tool. The sixth and seventh columns (Fault→Warn and Fault→Ok) report the percentage of failures in the fault detection process, which is the percentage of misclassification of faulty measurements, classified as being in a warning or a fault-free state, respectively, by the automatic tool.

Table 9.3. Performance of the validation tool

#Ch	Ok ↓ Ok	Ok ↓ Fault	Ok ↓ Warn	Fault ↓ Fault	Fault ↓ Warn	Fault ↓ Ok
1	92.4	3.4	4.2	95.5	2.7	1.8
2	90.0	6.2	3.8	95.2	1.6	3.2
3	94.4	5.6	0	72.9	19.4	7.7
4	90.9	5.6	3.5	98.1	0	1.9
5	85.4	6.7	7.9	86.0	8.3	5.7

Table 9.3 shows good performance in both validation and fault detection. Performance can be improved during the experimental life of the plant by re-tuning the fuzzy membership functions as more experimental data become available.

9.11 Basic Terminology in Fault Detection and Diagnosis

In the following, the terminology established by the IFAC Technical Committee SAFEPROCESS is reported (Isermann and Ballé, 1997; Omdahl, 1998).

1. **States and Signals**

   a. *Fault:* Unpermitted deviation of at least one characteristic property of the system.
   b. *Failure:* Permanent interruption of a system's ability to perform a required function under specified operating conditions.
   c. *Malfunction:* Intermittent irregularity in fulfillment of a system desired function.
   d. *Error:* Deviation between a computed value (of an output variable) and the true, specified or theoretically correct value.
   e. *Disturbance:* An unknown (and uncontrolled) input acting on a system.
   f. *Perturbation:* An input acting on a system which results in a temporary departure from steady state.
   g. *Residual:* Fault indicator, based on deviations between measurements and model equation based calculation.

h. *Symptom:* Change of an observable quantity from normal behaviour.

2. Functions

i. *Fault Detection:* Determination of faults present in a system and time of detection.
j. *Fault Isolation:* Determination of kind, location and time of detection of a fault by evaluating symptoms. Follows fault detection.
k. *Fault Identification:* Determination of the size and time-variant behaviour of a fault. Follows fault isolation.
l. *Fault Diagnosis:* Determination of kind, size, location and time of detection of a fault by evaluating symptoms. Follows fault detection. Includes fault detection, isolation, and identification.
m. *Monitoring:* A continuous real-time task of determining the possible conditions of a physical system, recognizing and indicating anomalies of the behaviour.
n. *Supervision:* Monitoring a physical system and taking appropriate actions to maintain the operation in the case of faults.
o. *Protection:* Means by which a potentially dangerous behaviour of the system is suppressed if possible, or means by which the consequences of a dangerous behaviour are avoided.

3. Models

p. *Quantitative model:* Use of static and dynamic relations among system variables and parameters in order to describe a system's behaviour in quantitative mathematical terms.
q. *Qualitative model:* Use of static and dynamic relations among system variables and parameters in order to describe a system's behaviour in qualitative terms such as causalities or if-then rules;
r. *Diagnostic model:* A set of static or dynamic relations which link specific input variables – the symptoms – to specific output variables – the faults.
s. *Analytical redundancy:* Use of two, but not necessarily identical ways to determine a quantity where one way uses a mathematical process model in analytical form.

4. System Properties

t. *Reliability:* Ability of a system to perform a required function under stated conditions, within a given scope, during a given period of time. Measure: MTTF = Mean Time To Failure. Also, MTTF=1/λ; where λ is the rate of failure (*i.e.* failures occurring in a time unit).

u. *Safety:* Ability of a system not to cause danger for persons or equipment or environment.
v. *Availability* Probability that a system or equipment will operate satisfactorily and effectively at any point of time:

$$A = \frac{\text{MTTF}}{\text{MTTF} + \text{MTTR}}$$

where MTTR is the Mean Time To Repair (MTTR=$1/\mu$; where μ is the rate of repair, *i.e.* the repairs made in a time unit).

9.12 Conclusions

The need to apply fault detection and diagnosis strategies to sensors is strongly felt in industrial applications. When applied to sensors, fault detection and diagnosis is called sensor validation. Although much effort has been devoted to set a general methodology for fault detection and diagnosis, it has been recognized that when attention is switched to industrial problems, the related solutions become more and more customized to the specific application. In this framework, soft sensors can be valuable tools to develop effective sensor validation tools. Nevertheless, in most cases, a successful sensor validation strategy cannot rely only on the development of an accurate soft sensor.

The applications presented in this chapter, developed in the nuclear fusion field, are based on a hybrid approach, which is able to conjugate the benefits of knowledge engineering with those of approximate reasoning, nonlinear black-box modeling, and optimization tools. Even though a general methodology cannot be set, suitable guidelines for an effective design are outlined in the chapter.

The tools presented are conceived in a modular structure, in which different validation strategies can be applied to cope with specific aspects of sensor validation. Subsequently, judgements about the measurement quality, computed through different techniques, can be finally merged to perform a global validation and fault classification. To this aim, soft computing techniques seem particularly suitable. In particular, artificial neural networks are useful tools to design nonlinear black-box models, whereas fuzzy logic is particularly suitable to incorporate linguistic knowledge about the phenomena involved, to represent both linguistic and numerical knowledge in a unified fashion, and to develop expert systems for fault diagnosis and classification. Moreover, thanks to their adaptation and learning ability, soft computing tools are useful to cope with those cases in which a lack of knowledge can be filled through the exploitation of experimental data.

It is worth noting that a hybrid approach does not exclude the presence of traditional fault detection and diagnosis techniques (*i.e.* techniques based on traditional signal processing and system analysis theory), which can be integrated as further modules in the validation tool.

This approach allows the designer to develop or update the various functional blocks separately, or to add new blocks if necessary. Moreover, through the use of a modular hybrid approach and soft computing tools, the following achievements can be highlighted:

- **Human expertise:** it is possible to transfer human knowledge about the phenomena involved to automatic validation tools, both for detection and classification purposes.
- **Flexibility:** if new kinds of fault appear, it is possible to detect and/or classify them just by modifying the rule set of fuzzy inference systems, training the models to provide additional features (symptoms), or adding extra estimation capability via signal processing, fuzzy analysis modules, and so on.
- **Adaptation:** when nonlinear black-box models (often based on neural networks) are used, if model performance worsens for some reason (typically because of totally new operational scenarios), it can easily be improved by an additional training session, to let the model auto-adapt to the new operational reality. Regular training sessions can be planned to keep models always at its best performance.
- **High degree of automation:** the approach proposed is particularly suitable to achieve a high degree of automation, strongly felt in plants where the huge amount of variables, the extreme variability of the experiments, the fast transients involved, and the primary need for safety are facts which strongly limit the ability of traditional methods of fault detection and classification.

Appendix A

Description of the Plants

A.1 Introduction

In this Appendix, real processes used as case studies in the various book chapters are described in some detail. Nevertheless, the tags of the variables involved in the following descriptions will be omitted for reasons of confidentiality. Care has been taken to keep to descriptions that will be as useful as possible for the reader to follow.

One set of processes described concerns a large refinery where soft sensors were required to monitor and control routine production. The final two case studies refer to two experimental nuclear fusion plants, where model-based sensor validation strategies were required to assess the quality of experimental observations made during research experiments.

A.2 Chimneys of a Refinery

The atmosphere contains a number of nitrogen oxides. Among these, nitric oxide (NO) and nitrogen dioxide (NO_2) are collectively given the name NO_x. The level of nitrogen oxides in general, and of NO_x in particular, in the atmosphere has been increasing in the last century, mainly due to human activities. Unfortunately, nitrogen oxides have a number of negative effects on air quality: they contribute to photochemical smog, cause reduction of visibility and acid rain and, last but not least important, they have a negative impact on human health.

The main source of NO_x is fossil fuel combustion, caused by human activities both in urban and industrial areas. However, it has a short lifetime in the atmosphere so that its effect tends to be regional, with greater concentrations in urban and industrial areas than in rural ones.

In past decades, a number of industrialized nations enforced more and more restrictive limits on NO_x emissions and, as a consequence, sophisticated monitoring strategies for NO_x emissions are now a matter of interest. It is also worth noting that they are included in the Kyoto treaty.

Stationary sources, such as power plants and refineries, significantly contribute to the emission of nitrogen oxides by means of chimney fumes produced by the combustion of residuals deriving from a number of processes. Hence, industries are deeply interested in the development of measurement and/or estimation strategies for these pollutants.

In particular, Italian laws establish a limit on the monthly average NO_x emission level, computed on the base of hourly recorded data. Moreover, the average value is considered valid if the minimum percentage of acquired hourly data is 80% or greater. In the event that such a percentage is not reached, *e.g.* due to failure of on-line analyzers used for NO_x measurements, Italian laws require the use of mathematical models that estimate their level on the basis of chimney inputs.

Liquid (called fuel oil, FO) and gaseous residuals (called fuel gas, FG), produced by a large number of processes, are burnt in a number of reactors and the resulting fumes are conveyed to the refinery chimneys, as schematically reported in Figure A.1, in order to minimize NO_x emission. In fact, it is generally accepted that conveying a large number of process residuals into big chimneys has a positive effect on the total amount of emitted NO_x.

NO_x emissions are measured using an on line analyzer, located as reported in Figure A.1, where it is indicated with an arrow. The measuring instrument is a gas chromatograph that measures the NO_x level with a sampling time $T_s = 1$ min.

To have comprehensive information about emitted pollutants, other chemicals are also monitored. In particular, measuring instrumentation is also used by the refinery to acquire data on:

- hash;
- sulfur oxides (SO_2);
- carbon oxide (CO).

Data produced by the analyzer are collected in a refinery database, marked with a corresponding validity flag, and subsequently used to compute the monthly average value. Due to the harsh environment, the analyzer is frequently off-line for scheduled maintenance and during these periods mathematical models need to be used to estimate the NO_x level.

Eighteen fuel gas and fuel oil flows considered in the models described in this book represent the inputs to the reactors.

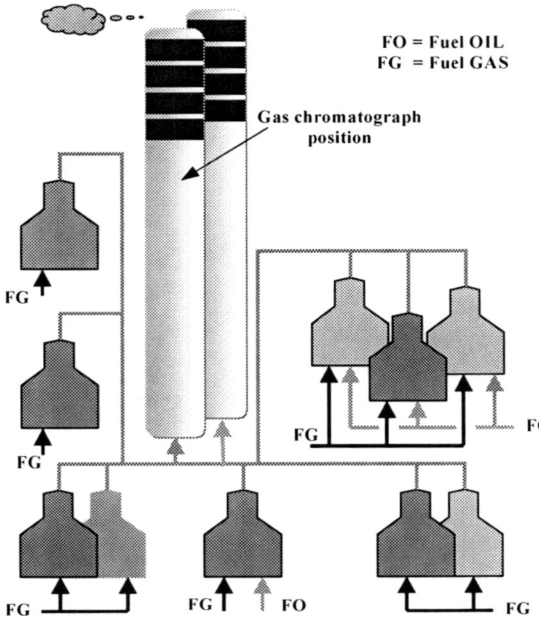

Figure A.1. A schematic view of chimneys used in the refinery to reduce the total amount of NO_x emission in the atmosphere. FG and FO from a number of refinery plants are burnt in reactors and the fumes obtained are conveyed to the chimneys. The arrow indicates the location of the on-line gas chromatograph used to acquire data on NO_x concentration.

Data considered in the design of soft sensors described here were obtained using records produced by the gas chromatograph during a period lasting about six months; they were collected in the plant database and used by plant operators for the estimation of mean monthly average emission values.

A.3 Debutanizer Column

The debutanizer column is part of a desulfuring and naphtha splitter plant, shown in Figure A.2, where two gray circles, A1 and A2, can be recognized. They represent the location of two gas chromatographs whose data were used to design soft sensors, while the two white circles, N1 and N2; indicate the points where soft sensors were required.

In particular, data acquired by the device A2, *i.e.*, the C4 (butane) content in the bottom flow to stock have been used in Chapter 6 as a study case for the design of the Soft Sensor named N2.

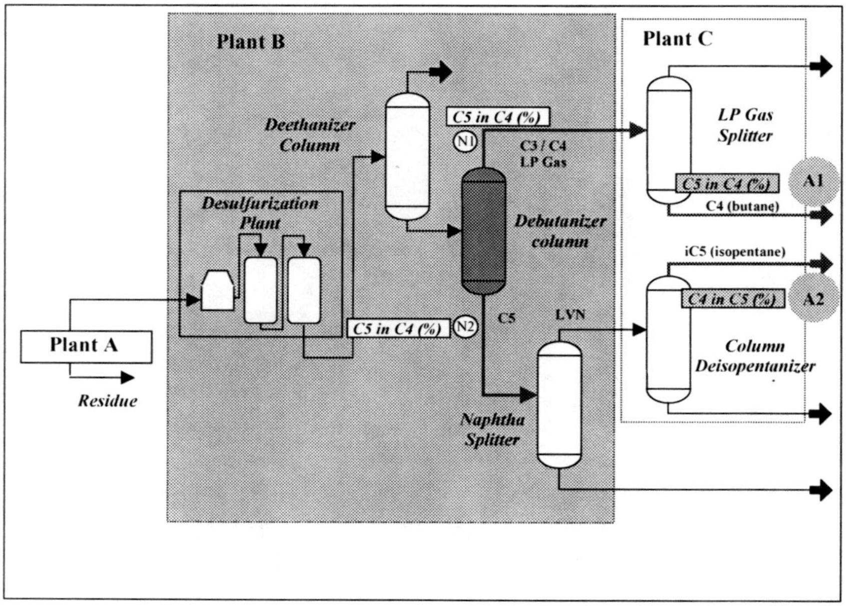

Figure A.2. Schematic view of the debutanizer column and connected plants. The location of two gas chromatographs (A1 and A2) is shown with two gray circles. N1 and N2 indicate the points where soft sensors were required

In the debutanizer column C3 (propane) and C4 (butane) are removed as overheads from the naphtha stream.

The debutanizer column is required to:

- ensure sufficient fractionation in the debutanizer;
- maximize the C5 (stabilized gasoline) content in the debutanizer overheads (LP gas splitter feed), while respecting the limit enforced by law;
- minimize the C4 (butane) content in the debutanizer bottoms (Naphtha splitter feed).

A detailed scheme of the debutanizer column is shown in Figure A.3.

A number of sensors, indicated with circles in Figure A.3, are installed on the plant to monitor product quality. The subset of sensors relevant to the application described, indicated with gray circles in Figure A.3, is listed in Table A.1, together with the corresponding description.

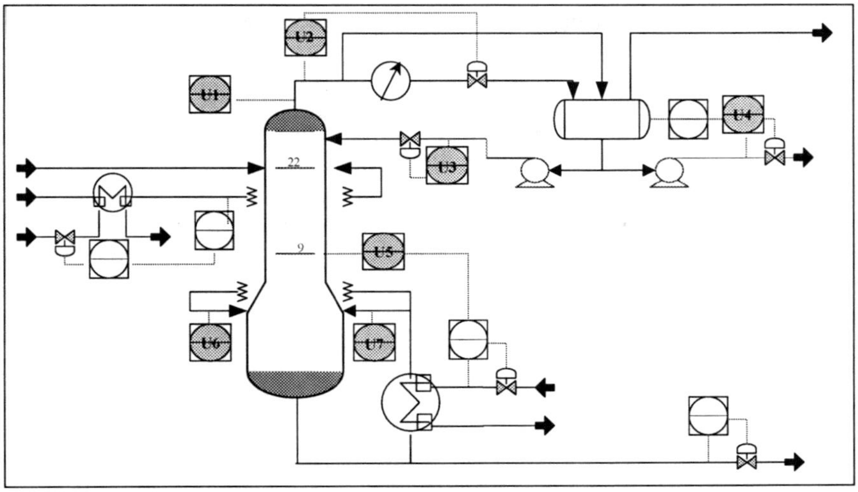

Figure A.3. Block scheme of the of the debutanizer column. Variables, as used in the case study described in Chapter 6, are indicated, along with the corresponding names, with gray circles. Open circles indicate variables, measured by instrumentation used by the refinery, but not used in the applications described

Table A.1. List of variables used in the design of soft sensors for the debutanizer column described in Chapter 6. Instrumentation location is reported in Figure A.3

Variable	Description
u_1	Top temperature
u_2	Top pressure
u_3	Reflux flow
u_4	Flow to next process
u_5	6th tray temperature
u_6	Bottom temperature
u_7	Bottom temperature

The C4 content in the debutanizer bottoms, *i.e,* the Soft Sensor output, is measured on the overheads of the deisopentanizer column, as can be observed in Figure A.2, where the location of the measuring device is indicated by the gray circle named A2. It measures the C4 content in the flow to stock that can all assumed to be coming out of the debutanizer bottoms.

A.4 Powerformer Unit

The powerformer unit used as a case study in Chapters 5 and 7 is shown schematically in Figure A.4.

The powerformer unit is designed to produce reformed gasoline with specified RON values. The RON value of the reformed gasoline, used to monitor the product quality and to control the powerforming process, is measured using a NIR analyzer. It receives as input the heavy virgin naphtha (HVN) flow from the Naphtha Splitter bottom that, combined with H_2, feeds the first of four cascaded reactors; furnaces between them are used to obtain the designed temperature profile during the catalytic process.

The output flow is a liquid high in octane number (RON) and rich in aromatic composites. Hydrogen, oil gas, and liquefied petroleum gas (LPG) are also obtained. The output flow feeds the de-ethanizer and debutanizer distillation columns.

Depending on the input flow, two different RON target values are defined. Also, the catalytic reactors need to be periodically regenerated and this phase is monitored on the basis of the four reactor temperature profiles. During the regeneration phases, the RON value drops below the desired level and flow is conveyed to an *off-spec* tank.

The process variables used in soft sensor design described in Chapters 5 and 7 are reported in Table A.2.

Table A.2. List of variables used in the design of soft sensors for the powerfomer unit described in Chapters 5 and 7

Variable	Description
u_1	RX1 Temperature
u_2	RX2 Temperature
u_3	RX3 Temperature
u_4	RX4 Temperature
u_5	Input flow
u_6	Top debutanizer pressure

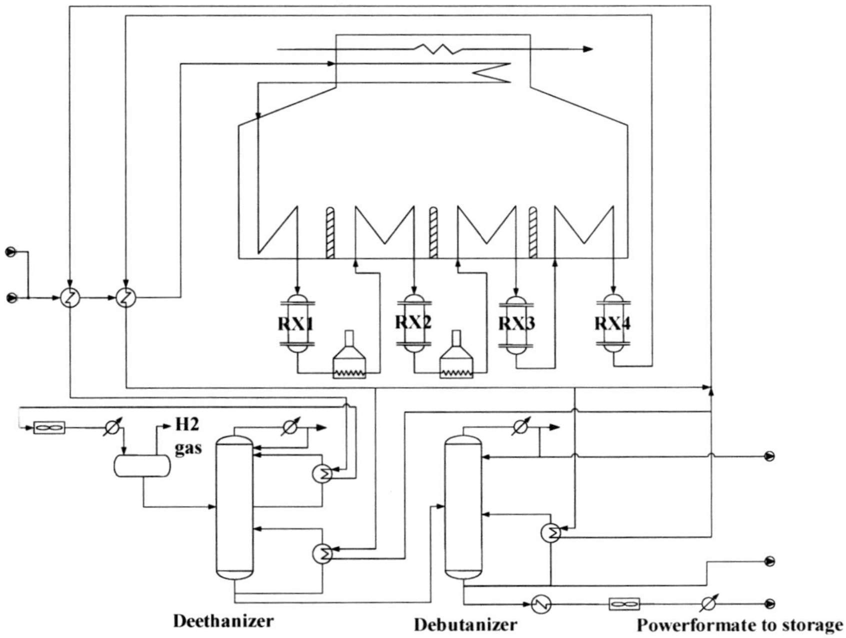

Figure A.4. A block scheme of the Powerformer Unit

A.5 Sulfur Recovery Unit

The sulfur recovery unit (SRU) removes environmental pollutants from acid gas streams before they are released into the atmosphere. Furthermore, elemental sulfur is recovered as a valuable by-product.

Acid gases are among the most dangerous air pollution factors and are one of the main causes of acid rain. Hydrogen sulfide is particularly dangerous because it prevents the cells of the human body from breathing.

The SRU takes in two kinds of acid gases as input. The first, rich in H_2S, called MEA gas, comes from the gas washing plants; the second, called SWS gas, rich in H_2S and NH_3 (ammonia), comes from the sour water stripping (SWS) plant. Acid gases are burnt in reactors, where H_2S is transformed into pure sulfur via a partial oxidation reaction with air. Gaseous combustion products from furnaces are cooled, causing the generation of liquid sulfur, which is collected in catch basins, and then passed through high temperature converters, where a further reaction leads to the formation of water vapor and sulfur. The remaining, non converted gas (less than 5%), is fed to the Maxisulfur plant for a final conversion phase. The final gas stream (tail gas) from the SRU contains residual H_2S and SO_2.

A simplified scheme for the SRU whose data have been used in the book is illustrated in Figure A.5. It is made up of a reaction furnace, which is divided into two separate combustion chambers.

The main chamber is fed with MEA gas, and combustion is regulated, in air deficiency, by supplying an adequate air flow (AIR_MEA). The secondary combustion chamber is mainly fed with SWS gas and a suitable air flow is provided (AIR_SWS).

The combustion of SWS gas occurs in a separate chamber with excess air, in order to prevent the formation of ammonium salts in the equipment, thereby giving rise to the generation of nitrogen and nitrogen oxides. The gas input flow in the secondary chamber is kept constant by adding some MEA gas (MEA_SPILLING) when the SWS gas flow is too low. Thus, an adequate air flow is added (MEA_SPILLING_AIR). Air flows are controlled by plant operators to guarantee a correct stoichiometric ratio in the tail gas. Control is improved by a closed-loop algorithm which regulates a further air flow (AIR_MEA_2) on the basis of analysis of the tail gas composition.

Air, which supplies oxygen for the reaction, is an important parameter in the conversion of H_2S, being responsible for the tail gas composition. In particular, an excessive air flow tends to increase the concentration of SO_2 with respect to H_2S, whereas a low air flow leads to the opposite situation.

On-line analyzers are used to measure the concentration of both hydrogen sulfide and sulfur dioxide in the tail gas of each sulfur line. The analyzers adopted are able to measure the quantity $[H_2S]$-$2[SO_2]$ (where the brackets indicate concentration), in order to monitor the performance of the conversion process and control the air-to-feed ratio in the SRU with the aim of improving the sulfur extraction process.

Hydrogen sulfide and sulfur dioxide frequently cause damage to sensors, which often have to be removed for maintenance. The design of soft sensors able to predict H_2S and SO_2 concentrations is therefore required, as in the case study reported in Chapter 5.

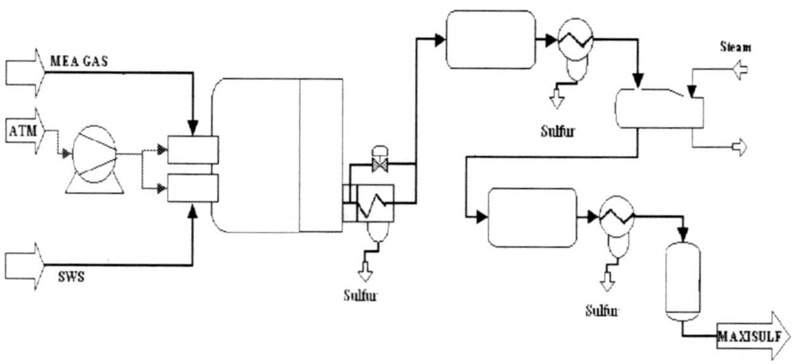

Figure A.5. Block scheme of the SRU

In order to predict the concentration of H_2S and SO_2 in the tail gas using soft sensors, the variables listed in Table A.3 were used.

Table A.3. List of variables used in the design of soft sensors for the SRU described in Chapter 5

Variable	Description
u_1	gas flow MEA_GAS
u_2	air flow AIR_MEA
u_3	secondary air flow AIR_MEA_2
u_4	gas flow in SWS zone (SWS_GAS_TOT=SWS_GAS+MEA_SPILLING)
u_5	air flow in SWS zone (AIR_SWS_TOT=AIR_SWS+MEA_SPILLING_AIR)

A.6 Nuclear Fusion Process: Working Principles of Tokamaks

In a nuclear fusion reaction, the nuclei of light elements (such as hydrogen) fuse together to form heavier ones, producing energy as a by-product. Nuclear fusion can be considered, in some sense, as the opposite of nuclear fission, a well-established technology in which energy is released by splitting heavy nuclei, such as uranium, in controlled chain reactions. The drawbacks of nuclear fission are related to the risk of nuclear meltdown with a large, uncontrolled release of energy, followed by heavy nuclear contamination. Moreover, fission reagents and by-products are highly radioactive and must be stored, handled, and disposed of with special care; in addition, they need to be kept under control for thousands of years. Unlike the case of fission, the loss of control in a fusion process naturally leads to the end of the reaction, without causing any disastrous nuclear meltdown; reagents and by-products can be safely stored and disposed of within a few decades. Fusion appears to offer many advantages over any other form of power production. Nevertheless, to control fusion for effective power production is an extremely challenging activity, involving scientists and engineers from all over the world. At present, only reactor prototypes exist, which are not designed to produce energy, but only have the aim of investigating various aspects of fusion technology.

Below, the basic principles underlying nuclear fusion are described, and the working principles of one of the most promising reactor prototypes, the tokamak, are given (Wesson, 1987). Finally, the measurement systems which are the subject of the sensor validation strategies presented in this book are described.

A.6.1 Nuclear Fusion

A typical fusion reaction can occur between two heavy isotopes of hydrogen, deuterium and tritium:

$$D + T \rightarrow {}^{4}He + n + \text{Energy} \tag{A.1}$$

Most of the energy released in this reaction is possessed by the high speed neutron, indicated in (A.1) with n. The remaining energy is carried by the helium nucleus ^{4}He, also called alpha-particle. In a fusion reactor, a lithium jacket or blanket around the reactor region would slow down the neutrons, converting their energy into heat. This heat could be extracted to raise steam for conventional electricity generation.

As deuterium is a common and readily separable component of water, there is a virtually inexhaustible supply in the oceans of the world. In contrast, tritium does not exist naturally in significant quantities, and must be manufactured. This can be done by exploiting chemical reactions between the neutrons formed during the fusion reactions and the lithium present in the blanket. As a consequence, although the fusion reaction occurs between deuterium and tritium, the consumables will be deuterium and lithium, as described by the following reactions, where ^{6}Li and ^{7}Li indicate heavier lithium isotopes:

$$\begin{aligned} &^6\mathrm{Li} + n \rightarrow \mathrm{T} + {}^4\mathrm{He} \\ &^7\mathrm{Li} + n \rightarrow \mathrm{T} + {}^4\mathrm{He} + n \end{aligned} \tag{A.2}$$

Experts agree that the reserves of lithium available are sufficient to enable world electricity generation using fusion reactors, to be maintained at present levels for several hundreds of years.

Under a fusion reaction, hydrogen isotopes change their state, becoming a gaseous mixture of ionized particles, which is called plasma. An important property of plasma, which makes it very different from a gas, is that it can be shaped and moved by magnetic fields. Plasma is often referred to as the fourth state of matter.

Fusion reactions can only take place if the nuclei are brought close to one another. However, this requirement is opposed by the fact that all nuclei carry a positive electric charge and therefore repel each other. By heating the gaseous fuels to very high temperatures, sufficient energy can be provided to the atoms to overcome the repulsive force and make them fuse together. In the deuterium–tritium reaction, temperatures in excess of 10^8 K are required, several times hotter than the center of the sun. Below 10^8 K, the D–T reaction rate falls off very rapidly: to one-tenth at 0.5×10^8 K, and 20000 times lower at 10^7 K. Concurrently, the plasma must be kept under very high pressure, in order to keep the particles very close to each other and increase the frequency at which they collide and consequently fuse.

In an effective reactor, the fusion reaction must obviously be self-sustained, *i.e.* more energy must be produced than that consumed to initiate and maintain the reaction.

A straightforward performance parameter for a reactor is Q, which is the ratio between output and consumed power. In a reactor for power production, the condition $Q>>1$ should obviously be achieved. $Q=1$ has already been achieved in experimental plants. In the so-called burning plasma, the high-temperature alpha-particles produced by the fusion reaction carry more energy than that supplied by external heating. This condition corresponds roughly to $Q>5$. When the alpha-particles can provide sufficient heat to self-sustain the reaction without external sources of heating, the ignition condition is achieved, which corresponds to $Q=\infty$.

Reactor power output depends on the square of the density of nuclei n_i, and on the volume of the gas. Power losses must also be kept to a minimum acceptable level by holding the hot gases in thermal isolation from their surroundings. The effectiveness of this isolation can be measured by the energy confinement time τ_E, which is the time taken for the system to cool down once all external forms of heating are switched off. Energy confinement time typically takes into account the capability of a reactor to achieve a steady-state operational condition.

In a fusion reactor, the values of density, energy confinement time, and temperature must be such that their product $n_i \cdot \tau_E \cdot T_i$ exceeds 5×10^{21} m$^{-3}\cdot$s$\cdot$keV. Typical values for the parameters that must be attained simultaneously for a reactor are:

- Central ion temperature T_i=10–20 keV;
- Central ion density $n_i=2.5\times10^{20}$ m^{-3};
- Energy confinement time τ_E=1–2 s.

Extremely high temperature and density are mandatory elements to obtain a self-sustained plasma. At the same time, they are the main causes of instability: the more temperature and density increase, the more unstable is the plasma. Consequently, the fundamental challenge in fusion technology is to discover adequate methods for confining and heating the plasma, without undergoing destructive instabilities.

Under the name of plasma confinement we refer to all the techniques able to maintain a plasma within a prescribed volume. Because of its nature, plasma cannot be contained in a volume in the usual sense, that is to say, plasma cannot physically touch the walls of the volume in which it is created. The interaction with the walls, in fact, leads to plasma destruction and, consequently, to an abrupt ending of the reaction.

A plasma can be confined through three different physical principles:

- *Gravitational confinement.* This is the way the sun confines plasma around itself. Plasma is confined because of extremely high gravitational forces.
- *Inertial confinement.* With this technique, hydrogen gases are compressed through a controlled implosion. The consequent inertia is able to keep particles close enough to make the reaction occur.
- *Magnetic confinement.* As the plasma is a mixture of charged (ionised) particles, magnetic fields can be exploited to maintain, shape, and control plasma.

Magnetic confinement is one of the most promising confinement techniques. A very effective structure in which plasma can be magnetically confined is the *tokamak.* The next section deals with the basic working principles of tokamaks.

A.6.2 Tokamak Working Principles

A.6.2.1 Main Phases of a Fusion Reaction in Tokamaks

Tokamak is originally a Russian acronym for toroidal chamber and magnetic coil. As the name implies, in a tokamak plasma is confined within a toroidal structure, by means of suitable magnetic fields generated by electric currents flowing in coils. There are usually several coils placed around the plasma, with the purpose of generating, shaping, heating plasma, and driving current in the plasma column.

Several tokamaks have been installed all over the world, each of them with its peculiar characteristics. Nevertheless, all of them share the same working principles, which are described below.

All tokamaks are pulsed devices, in which plasma is created and maintained for a period of time ranging from a few seconds to a few minutes. However, it has not been envisioned yet whether a power reactor will be working in a real steady-state or in a pulsed fashion with very long pulses.

Each experiment on a tokamak, called pulse, or shot, or discharge, consists mainly of the same sequence of events. To illustrate the working principles of a tokamak pulse, we refer to the JET machine, the largest fusion facility in the world, managed by EFDA, located in Culham, Oxfordshire, UK.

Plasma is confined in a toroidal chamber, the vacuum vessel. The JET vacuum vessel is made up of eight identical sectors or octants. Each octant is composed of thick rigid box sections and bellows. Into the vacuum vessel, a vacuum condition of 10^{-9} mbar has been achieved through strict manufacturing and cleanliness procedures. The vessel is bakable up to 500°C under vacuum, to remove gas and impurities adsorbed on the inner wall surface.

In a tokamak, two main magnetic fields must be distinguished: the toroidal field is a magnetic field, acting along the toroidal axis of the vacuum vessel. All the magnetic fields acting on a plane which is orthogonal to the toroidal axis of the vessel are referred to as poloidal fields.

The toroidal field is generated by a set of 32 identical, D-shaped, conventional copper wound, water-cooled coils. The toroidal coils are wound around the minor circumference of the torus and equally spaced around the machine. At the beginning of the experiment, prior to plasma formation, the toroidal field is brought to a constant value, in order to confine the plasma when it is initially created. After the toroidal field has reached a steady value, either hydrogen or deuterium is puffed into the vacuum vessel.

Simultaneously with the gas puffing, the current in the inner poloidal coil located at the center of the torus is brought up to its maximum value in preparation for pulse initiation. Subsequently, its current is driven down very quickly to produce a large electric field able to provide the energy needed for plasma creation. As plasma is made up by charged particles free to move, it is actually a conductor. Therefore, a transformer effect is established, in which the inner poloidal coil is the primary one and the plasma column is the secondary. At JET, this effect is enhanced by a transformer iron circuit, made up of eight limbed transformer cores surrounding the vessel. The transformer effect causes an electric current to flow into the plasma along the toroidal axis (called plasma current, I_p), by means of the opposing flows of oppositely charged particles. The combination of toroidal and poloidal fields on the one hand, with the plasma current on the other, produces a helical magnetic field which, in principle, is able to keep the plasma away from the vessel walls and confined in a toroidal shape.

The initial plasma heating is produced by the plasma current, as the collisions of the plasma electrons and ions make the plasma resistive. Therefore, this form of plasma heating is called ohmic heating. The current in the inner poloidal coil is now regulated in order to maintain the plasma current at a target flat-top value. Unfortunately, the ohmic heating effect is reduced as the electrical resistance of the plasma decreases with increasing temperature. Therefore, the plasma must be supplied with additional heating by several means. Two common ways of achieving this are the injection of neutral particles beams, commonly referred to as neutral beam heating or the emission of high-power radiofrequency waves.

The three phases, during which the plasma current I_p increases, is kept constant, and decreases, are often referred to, respectively, as ramp-up, flat-top, and ramp-down phases of the plasma.

In addition to the inner poloidal coil, outer poloidal field coils are installed. Their task is to control the shape and position of the plasma. The inner poloidal coil is actually divided into two sections. The outer sides of the coil are used exclusively for driving current into the plasma and for ohmic heating, whereas the central portion of the coil works together with the outer coils in shaping and controlling the position of the plasma column. Plasma position and shape are, in fact, unstable and the helical field arising as a combination of the toroidal field, the main poloidal field and the plasma current is not sufficient to reach a stable condition. The greatest instability of the plasma is along the vertical direction, and control of the vertical position of the plasma is a challenging task.

A.6.2.2 Plasma–Surface Interactions: Limiter and Divertor Configuration

A fusion reaction must occur in an environment free from impurities of any kind. Since the very first experimental studies on tokamaks, it has been found that a vessel in strict vacuum condition is not sufficient to guarantee pure plasma. Impurities can arise from the interaction of plasma with components located at the interior of the vessel, and they can also be released within the reaction. In particular, the helium by-product of the fusion reaction (also called helium ash) can interfere with subsequent fusion reactions. All impurities tend to remain in the plasma for a finite time before leaking out. When impurities leave the plasma, they are still charged and tend to follow magnetic field lines.

A first strategy for coping with impurities is to limit the plasma confinement region by inserting a material structure, called a limiter, into the vacuum vessel. The limiter intercepts a fraction of the magnetic lines and creates a separation between the plasma and the interior plasma-facing wall of the vessel (also called the first wall). As a consequence, the last closed flux surface which confines the plasma is kept separated from the first wall. Therefore, the limiter acts by protecting the chamber wall from plasma bombardment and helps in defining the edge of the plasma. The high energy impurity particles which leave the plasma collide with the limiter and can dislodge atomic impurities (*i.e.* the process of sputtering). Particular care must therefore be adopted in the choice of limiter material, and pumping devices (such as cryocondensation pumps) must be installed to remove the impurities created by the sputtering process.

Alternatively, it is possible to exploit the fact that impurities are electrically charged and tend to follow a magnetic field line. Consequently, they can be removed by designing adequate divertor regions, in which nonhydrogen particles can be diverted away from the plasma and subsequently cooled down and pumped out from the vacuum vessel. A tokamak endowed with divertors has additional coils (*i.e.* divertor coils), able to create a drop-shaped plasma, in such a way that the external magnetic lines are guided away from the main plasma and collide with a collector plate. With this magnetic configuration, impurities leave the plasma and subsequently strike against the first wall at fixed locations called strike points.

A.6.2.3 Plasma Disruptions

Fusion experiments performed in tokamaks are intrinsically unstable. Large-scale plasma instabilities, which in most cases are the main causes of an abrupt end of the pulse in a tokamak are called plasma disruptions. These are fast events in which most of the plasma thermal energy is rapidly lost. Most disruptions, called major disruptions, lead to almost instantaneous termination of the plasma current, whereas, more rarely, minor disruptions can occur, in which the energy loss is less important and the experiment can be recovered.

Plasma disruptions can be preceded by a loss of vertical stability. This is the most frequent case in tokamaks producing vertically elongated (D-shaped) plasmas, which are intrinsically unstable along the vertical direction, like that produced in the JET apparatus. In this case, the event which triggers the disruption is called a vertical displacement event (VDE), and occurs when vertical position control is lost. Then, the position instability tends to grow, the plasma strikes the internal surface of the vessel, eventually producing a large-scale instability which causes the loss of all plasma thermal energy. Consequently, the plasma becomes too cold and resistive, the transformer effect induced by the main poloidal field is no longer able to sustain the plasma current, and the plasma current is suddenly terminated. Disruptions always cause important thermal and mechanical stress on the tokamak structure, which must be carefully monitored. In view of this, the JET machine has been endowed with a measurement system, called the machine diagnostic system (MDS). A subset of MDS sensors, in particular a set of 32 strain gauges located at the vertical restraints of the vessel, is the subject of the first sensor validation strategy presented earlier in this book. The measurement systems and the background needed to develop the related sensor validation strategy is introduced in the next section.

A.7 Machine Diagnostic System at JET and the Monitoring of Mechanical Stresses Under Plasma Disruptions

A.7.1 The MDS Measurement System

The MDS is a measurement system developed at JET by the Magnet and Power Suppply Division (Marchese *et al.*, 1997). It was installed during the 1995/96 JET shutdown and is devoted to performing a large set of mechanical measurements, such as forces, displacements, accelerations, and pressures, along the mechanical structure of the JET machine. At present, it consists of 512 measurement channels. All the installed probes are passive, due to the high neutron fluxes expected during operation under active conditions, and the noise induced by the high magnetic fields is reduced by means of a carrier at 5 kHz, followed by demodulation and filtering.

The top and bottom main vertical port (MVP) restraints (lockable brakes), equipped with 64 strain gauges, support most of the vertical force acting on the vacuum vessel during plasma disruptions caused by VDEs. The axial movement of the brakes is measured with 32 linear variable resistors (LVRs). Up to 62 LVRs are used to monitor a wide set of radial, vertical and tangential movements of vertical ports, horizontal ports, and inner walls.

The lateral restraints of the main horizontal ports (MHPs), recently introduced to reduce the vessel's sideways displacements, are monitored by 16 pressure gauges, together with four triaxial accelerometers, with a measurement range of 0 to 50 g ($g=9.81\ m/s^2$), installed at the MHPs of octants, 2, 4, 6, and 8.

The measured data are sent to a host computer via a dedicated network. Measurements are performed at a sampling rate of 2.5 kHz, and are stored, after smoothing and filtering, at different rates, according to the different operational status of the machine. Three sample ratings are used:

- *Fast* (2.5 kHz). This rate is used when either a disruption or a major plasma instability or misbehavior occurs. Related data are stored in a database called *Jet Pulse File* (JPF) for about 800 ms.
- *Slow* (25 Hz). This rate is used during the active phase of the pulse (from ramp-up to ramp-down), lasting about one minute.
- *Continuous* (0.25 Hz). This rate is used to evaluate statical properties of the experiment. Related data are continuosly stored.

The MDS data, flowing through a VME bus, are accessible in real time via a PC equipped with a double Pentium 300 MHz board, a VME-PCI adapter, and the LabView™ software package. Both thresholding and the algorithm based on physical redundancy described in Section 9.9.2 are continuously run on this PC and duty officers are warned via e-mail of the occurrence of detected sensor faults.

On the occurrence of a disruption, the validation algorithms described previously do not allow one to validate the vertical stress measurements, performed at fast rate by the set of 32 strain gauges installed on the bottom supports of the vacuum vessel. In view of this, the validation system described in Section 9.9.3 was developed.

A.7.2 Disruptions and Mechanical Stresses

With a disruption caused by a VDE, high mechanical and thermal stresses are produced on the machine structure (Buzio *et al.*, 1996). Disruption-induced loads are characterized by radial and vertical components of several millions of newtons, with typical time scales ranging from 20 to 50 ms. Since the installation of additional restraining rings (1989), the vessel has become more rigid with respect to radial axisymmetric forces and the most important mechanical loads are now due essentially to vertical forces. Because of the particular configuration of the supports, vertical forces exert a torque around a rotation centre, generating a vessel axisymmetric oscillating motion around a centre, called rocking motion, with a frequency of 14 Hz. Experts summarize the force history into the following phases (Buzio *et al.*, 1996):

1. initial steady state phase, during pulse flat-top, due to the interaction between plasma and the divertor coils;
2. upward force, due to eddy and halo currents induced by the plasma vertical instability;
3. large downward swing, due to the plasma current quench and to the currents induced in the divertor coils;
4. rocking motion at 14 Hz, caused by the fact that the vertical forces exert a torque around a rotation centre located near the root of the main vertical ports of the vessel, in the inner side.

The downwards swing occurring in phase 3 is the most important, with regard to the fatigue life of the vessel, considering that the machine is now operating above its original design goals. This swing is estimated by computing the so called F-number. This is a simple nonlinear function of seven typical currents, that are part of the settings of the experiments. They are:

1. The plasma current I_p
2. The current in the inner poloidal coil, that provides the energy to create the plasma by a transformer effect in which the inner poloidal coil is the primary and the plasma is the secondary, I_{pfx}
3. The current in the plasma shaping circuit (shaping current), multiplied by the effective number of turns in the shaping circuit, $N_{sh}\ I_{sh}$
4. The four divertor coil currents, I_{D1}, I_{D2}, I_{D3} and I_{D4}.

A good estimate of the downward swing is obtained by evaluating the F-number formula at a time known as *STIME*, 200 ms before the detected disruption time. This choice is adopted in order to avoid considering corrective actions, which are performed on the relevant currents by the plasma position control circuit in an attempt to recover from the disruption, that actually modifies the natural trend of the experiment. Consequently, the currents measured at *STIME* reflect the effective current values at the disruption more than the current values taken at the actual disruption time. For this reason, the input of the neural network described in Section 9.9.3 consists of the currents evaluated at *STIME*.

A.8 Interferometry-based Measurement System for Plasma Density at FTU

The measurement system described above is useful to investigate the mechanical stresses of the vessel. This issue is extremely helpful for engineers, who are concerned with keeping the plant in good and safe working condition.

Physicists, instead, are interested in the study of many aspects of the reactions. Many other measurement systems must therefore be developed in order to investigate all the physical phenomena involved in the reaction. These kinds of measurement systems are called *diagnostics* (Hutchinson, 1990) in the fusion community. However, we prefer not to use this term in this book, as when we refer to *diagnosis* it has a more extensive meaning (see Chapter 9). Diagnostics can be used in nuclear fusion for studying problems in five main areas:

1. Methods of setting up stable plasmas.
2. Determination of important plasma parameters such as energy, density, temperature, and particle confinement time.
3. Study of additional heating techniques.
4. Study and control of plasma impurities.
5. Control of plasma shape and position.

Plasma density measurement is of fundamental importance. To measure this quantity, one of the most widely used techniques is based on laser interferometry. Lasers are exploited as good diagnostic tools in plasma studies, as they are sufficiently bright and monochromatic to compete with the self-emission of the plasma in a narrow band; their use does not require electrodes, probes, or other protuberances inside the vessel, and they offer good spatial and temporal resolution, due to their natural collimated beams and the short duration of their pulses. A laser beam can interact with a plasma producing different modes of interactions: scattering, absorption, reflection, refraction and transmission. Many measurement systems can be designed and implemented by exploiting each of the basic interaction modes.

In a laser interferometer, the plasma electron density is measured by exploiting the change in the refractive index of the laser caused by the presence of the electron gas (*i.e.* the plasma). In a high temperature plasma, the refractive index is dominated by the electron contribution, as the ion to electron mass ratio is very large. In a plasma without external magnetic field, or collisions, the solution of the wave equation gives the following dispersion relation for the wave number k

$$k^2 = \frac{\omega_0^2}{c^2} - \frac{\omega_{pe}^2}{c^2} \tag{A.3}$$

where c is the light speed, ω_0 is the laser frequency, and ω_{pe} is the plasma frequency, which can be expressed as

$$\omega_{pe} = \left(\frac{n_e e^2}{\varepsilon_0 m_e} \right)^{\frac{1}{2}} \tag{A.4}$$

where e is the electron charge, m_e is the electron mass, n_e the plasma electron density and ε_0 is the vacuum permittivity. Generally $\omega_0 >> \omega_{pe}$. The refractive index, μ, can be derived from equation (A.3) as

$$\mu - 1 \cong -\frac{e^2}{2\varepsilon_0 m_e} \frac{n_e}{\omega_0^2} = -\left(\frac{e^2}{8\pi^2 \varepsilon_0 c^2 m_e} \right) n_e \lambda_0^2 \tag{A.5}$$

where λ_0 is the laser wavelength. As can clearly be seen from Equation A.5, the refractive index of a laser interacting with a plasma is a function of the plasma electron density. Because of the change in the reflective index, when a laser beam is transmitted through a plasma along a line l, it undergoes a phase shift $\Delta\phi$ which is function of the refractive index and, consequently, of the plasma electron density, as

$$\Delta\phi = \frac{2\pi}{\lambda_0} \int_l [1 - \mu(z)] dz = r_e \lambda_0 \int_l n_e(z) dz \tag{A.6}$$

where r_e is the classical electron radius and z the line coordinate.

The interferometer installed at FTU is based on the Mach-Zehnder type (Wesson, 1987). The laser adopted is based on a deuterium–carbonium–nitrogen (DCN) gas. In the interferometer, a probing beam is transmitted through the plasma, and the phase change is revealed by comparison with an outside reference beam. The measurement of the phase shift determines the integrated density along the chord through which the laser interacted with the plasma. The phase difference is counted by a high-frequency electronic clock, called fringe counter, and when the value 2π, *i.e.* one fringe, is reached the counter reverts to zero. A typical fault generated in an interferometer is fringe skip, which consists of missing whole fringes in the fringe counts. This leads to measurements affected by step-like discontinuities. After counting, fringes are converted into absolute line densities. In order to obtain the radial distribution of the plasma density, several measuring channels are used at varying distance from the plasma centre. A mathematical inversion procedure (Abel inversion) can then be used to find the density as a function of plasma radius. At FTU, a five-channel interferometer has been installed to measure plasma electron density. It performs the plasma line density measurement along five parallel lines.

Appendix B

Structured References

B.1 Theoretical Contributions

B.1.1 Books

Ahlberg JH, Nilson EN, Walsh JL, (1967) The theory of splines and their applications. Academic Press, New York

Barlow RE, Proschan F, (1975) Statistical theory of reliability and life testing. Holt, Rinehart and Winston, Inc.

Bezdek JC, (1981) Pattern recognition with fuzzy objective function algorithms. Plenum Press, New York

Bishop CM, (1995) Neural networks for pattern recognition. Clarendon Press, Oxford, UK

Chen J, Patton R, (1999) Robust model-based fault diagnosis for dynamic systems. Kluwer, Boston

Chiang LH, Russell EL, Braatz RD, (2001) Fault detection and diagnosis in industrial systems. Advanced Books in Control and Signal Processing Series, Springer-Verlag, London

Combs Clide F Jr, (1999) Electronic instrument handbook, McGraw-Hill Handbooks: 45.1–45.24

Fortuna L, Rizzotto G, Lavorgna M, Nunnari G, Xibilia MG, Caponetto R, (2001) Soft computing: new trends and applications. Advanced Textbooks in Control and Signal Processing, Springer

Gertler J, (1998) Fault detection and diagnosis in engineering systems. Marcel Dekker, New York

Guidorzi R, (2003) Multivariable system identification. Bonomia University Press, Bologna

Gupta MM, Sinha NK, (2000) soft computing and intelligent systems: theory and applications. Academic Press Series in Engineering, San Diego CA

Haykin S, (1999) Neural networks: a comprehensive foundation, 2nd edn. Prentice Hall

Himmelblau DM, (1978) Fault detection and diagnosis in chemical and petrochemical processes. Elsevier, Amsterdam

Isermann R, (2006) Fault-diagnosis systems, an introduction from fault detection to fault tolerance. Springer-Verlag Berlin Heidelberg

Isermann R, (2006b) Fault diagnosis of technical processes – applications. Springer-Verlag Berlin Heidelberg

Jang JR, Sun C, Mizutani E, (1997) Neuro fuzzy and soft computing: a computational approach to learning and machine intelligence. Prentice Hall

Kreyszig E, (1999) Advanced engineering mathematics: international edition. John Wiley & Sons Ltd

Ljung L, (1999) System identification: theory for the user, 2nd edn. T. Kailath, Ed. Englewood Cliffs, Prentice Hall

Looney CG, (1997) Pattern recognition using neural networks – theory and algorithms for engineers and scientists. Oxford University Press, Oxford, UK

NIST/SEMATECH (2005) e-Handbook of Statistical Methods, http://www.itl.nist.gov/div898/handbook/

Nørgaard M, Ravn O, Poulsen NK, Hansen LK, (2000) Neural networks for modelling and control of dynamic system. Springer, London

Omidvar O, Elliot DL, (1997) Neural systems for control. Academic Press, San Diego, CA, USA.

Oppenheim AV, Schafer RW, (1989) Discrete-time signal processing. Prentice-Hall

Patton R, Frank P, Clark R, (1989) Fault diagnosis in dynamic systems. Prentice Hall, Hertfordshire, UK

Patton R, Frank P, Clark P, (eds) (2000) Issues of fault diagnosis for dynamic systems. Springer, New York

Pau LF, (1981) Failure diagnosis and performance monitoring. Marcel Dekker, New York

Simani S, Fantuzzi C, Patton RJ, (2002) Model-based fault diagnosis in dynamic systems using identification techniques. Advances in Industrial Control Series, Springer, London

B.1.2 Data Collection and Filtering, Effect of Missing Data

Chiang LH, Pell RJ, Seasholtz MB, (2003) Exploring process data with the use of robust outlier detection algorithms. Journal of Process Control 13:437–449

Englund C, Verikas A, (2005) A hybrid approach to outlier detection in the offset lithographic printing process. Engineering Applications of Artificial Intelligence. 18:759–768

Lin B, Recke B, Renaudat P, Knudsen J, Jørgensen SB, (2005) Robust statistics for soft sensor development in cement kiln. 16th IFAC World Congress, Prague

Lopes VV, Menezes JC, (2005) Inferential sensor design in the presence of missing data: a case study. Chemometrics and Intelligent Laboratory System 78:1–10

Nelson PRC, Taylor PA, MacGragor JF, (1996) Missing data methods in PCA and PLS: Score calculations with incomplete observations. Chemometrics and Intelligent Laboratory Systems 35(1):45–65

Pearson RK, (2002) Outliers in process modeling and identification. IEEE Transactions on Control Systems Technology 20(1):55–63

Spinelli W, Piroddi L, Lovera M, (2005) On the role of pre-filtering in nonlinear system identification. 16th IFAC World Congress, Prague

Thornhill NF, Shoukat Choudhury MAA, Shah SL, (2004) The impact of compression on data-driven process analyses. Journal of Process Control 14:389–398

Warne K, Prasad G, Rezvani S, Maguire L, (2004) Statistical and computational intelligence techniques for inferential model development: a comparative evaluation and a novel proposition for fusion. Engineering Applications of Artificial Intelligence 17:871–930

B.1.3 Variables and Model Structure Selection

Albazzaz H, Wang XZ, (2006) Historical data analysis based on plots of independent and parallel coordinates and statistical control limits. Journal of Process Control 16:103–114

Baffi G, Martin EB, Morris AJ, (1999) Non-linear projection to latent structures revisited (the neural network PLS algorithm). Computers and Chemical Engineering 23:1293–1307

Bomberger JD, Seborg DE, (1998) Determination of model order for NARX models directly from input-output data. Journal of Process Control 8:459–468

Feil B, Abonyi J, Szeifert F, (2004) Model order selection of nonlinear input-output models – a clustering based approach. Journal of Process Control 14:593–602

Flynn D, Ritchie J, Cregan M, (2005) Data mining techniques applied to power plant performance monitoring. 16th IFAC World Congress, Prague

Haber R, Unbehahuen H, (1990) Structure identification of nonlinear dynamic systems – a survey on input/output approaches. Automatica 26(4):651–677

Hashem S, (1997) Optimal linear combinations of neural networks. Neural Networks 10(4):599–614

He X, Asada H, (1993) A new method for identifying orders of input-output models for nonlinear dynamic systems. Proc. of American Control Conference, San Francisco CA, 3:2520–2522

Komulainen T, Sourander M, Jamsa-Jounela SL, (2004) An online application of dynamic PLS to a dearomatization process. Computers and Chemical Engineering 28:2611–2619

Lang ZQ, Futterer M, Billings SA, (2005) The identification of a class of nonlinear systems using a correlation analysis approach. 16th IFAC World Congress, Prague

Lee JM, Yoo C, Lee IB, (2004) Statistical process monitoring with independent component analysis. Journal of Process Control 14:467–485

Lind I, (2005) Nonlinear structure identification with linear least squares and ANOVA. 16th IFAC World Congress, Prague

Lind I, Ljung L, (2005) Regressor selection with the analysis of variance method. Automatica 41:693–700

Liu J, (2005) On line soft sensor for polyethylene process with multiple production grades. 16th IFAC World Congress, Prague

Mendes E, Billings SA, (2001) An alternative solution to the model structure selection problem. IEEE Transactions on Systems Man and Cybernetics, Part A-System and Humans 31:597–608

Nagai EY, Arruda LVR, (2005) Soft sensor based on fuzzy model identification. 16th IFAC World Congress, Prague

Rallo R, Ferre-Giné J, Arena A, Giralt F, (2002) Neural virtual sensor for the inferential prediction of product quality from process variables. Computers and Chemical Engineering 26:1735–1754

Sridhar DV, Bartlett EB, Seagrave RC, (1999) An information theoretic approach for combining neural network process models. Neural Networks 12:915–926

Takagi T, Sugeno M, (1985) Fuzzy identification of systems and its applications to modelling and control. IEEE Transactions on Systems, Man, and Cybernetics 5(3):116–132.

Warne K, Prasad G, Rezvani S, Maguire L, (2004) Statistical and computational intelligence techniques for inferential model development: a comparative evaluation and a novel proposition for fusion. Engineering Applications of Artificial Intelligence 17:871–930

Zhang H, Zouaoui Z, Lennox B, (2005) A comparative study of soft-sensing methods for fed-batch fermentation processes. 16th IFAC World Congress, Prague

B.1.4 Model Identification

Araúzo-Bravo MJ, Cano-Izquierdo JM, Gómez-Sánchez EG, López-Nieto MJ, Dimitriadis YA, López-Coronado J, (2004) Automatization of a penicillin production process with soft sensors and a adaptive controller based on neuro fuzzy systems. Control Engineering Practice 12:1073–1090

Chen S, Billings SA, (1989) Representations of non-linear system: the NARMAX model. International Journal of Control 49:1012–1032

Feng S, Chen J, Tu XY, (2005) Nonlinear system identification with shortage of input–output data. 16th IFAC World Congress, Prague

Juditsky A, Hjalmarsson H, Benveniste A, Delyon B, Ljung L, Sjoberg J, Zhang Q, (1995) Nonlinear black-box models in system identification: mathematical foundations. Automatica 31:1725–1750

Kahrs O, Brendel M, Marquardt W, (2005) Incremental identification of NARX models by sparse grid approximation. 16th IFAC World Congress, Prague

Li K, Peng J, Irwin GW, Piroddi L, Spinelli W, (2005) Estimation of NO_x emissions in thermal power plants using eng-genes neural networks. 16th IFAC World Congress, Prague

Liu J, (2005) On line soft sensor for polyethylene process with multiple production grades. 16th IFAC World Congress, Prague

Luo JX, Shao HH, (2006) Developing soft sensors using hybrid soft computing methodology: a neurofuzzy system based on rough set theory and genetic algorithms. Soft Computing 10:54–60

Nagai EY, Arruda LVR, (2005) Soft sensor based on fuzzy model identification. 16th IFAC World Congress, Prague

Narendra KS, Parthasarathy K, (1990) Identification and control of dynamical systems using neural networks. IEEE Transactions on Neural Networks 1(1):4–27

Park S, Han C, (2000) A nonlinear soft sensor based on multivariate smoothing procedure for quality estimation in distillation columns. Computers and Chemical Engineering 24:871–877

Rallo R, Ferre-Giné J, Arena A, Giralt F, (2002) Neural virtual sensor for the inferential prediction of product quality from process variables. Computers and Chemical Engineering 26:1735–1754

Sjoberg J, Hjalmarsson H, Ljung L, (1994) Neural networks in system identification. 10th IFAC System Identification-Symposium, Cophenagen

Sjoberg J, Zhang Q, Ljung L, Benveniste A, Delyon B, Glorennec PY, Hjalmarsson H, Juditsky A, (1995) Nonlinear black-box modeling in system identification: a unified overview. Automatica 31:1691–1724

Yan W, Shao H, Wang X, (2004) Soft sensing modeling based on support vector machine and Bayesian model selection. Computers and Chemical Engineering 28:1489–1498

B.1.5 Model Validation

Billings SA, Jamaluddin HB, Chen S, (1992) Properties of neural networks with applications to modelling non-linear dynamical system. International Journal of Control 55:193–224

Billings SA, Voon WSF, (1991) Correlation based model validity tests for nonlinear models. International Journal of Control 54:157–194

Chen S, Billings SA, Grant PM, (1990) Non-linear system identification using neural networks. International Journal of Control 51:1191–1214

Dadhe K, Engell S, (2005) Assessing the predictions of dynamic neural networks. 16th IFAC World Congress, Prague

Masson MH, Canu S, Grandvalet Y, Lynggaard-Jensen A, (1999) Software sensor design based on empirical data. Ecological Modelling 120:131–139

Mendes EMAM, Billings SA, (2001) An alternative solution to the model structure selection problem. IEEE Transactions on Systems Man and Cybernetics, Part A-System and Humans 31:597–608

Papadopoulos G, Edwards PJ, Murray AF, (2001) Confidence estimation methods for neural networks: a practical comparison. IEEE Transactions on Neural Networks 12(6):1456–1464

Stoica P, Södersröm T, (1989) System identification. Englewood Cliffs, NJ, Prentice Hall

B.1.6 Fault Detection and Diagnosis, Sensor Validation

B.1.6.1 Survey works

Frank PM, (1990) Fault diagnosis in dynamic systems using analytical and knowledge-based redundancy – a survey and some new results. Automatica 26(3):459–474

Gertler J, (1988) Survey of model-based failure detection and isolation in complex plants. IEEE System and Control Magazine, December 1988:3–17

Henry MP, Clarke DW, (1993) The self validating sensor: rationale, definitions, and examples. Control Engineering Practice 1(4):585–610

Isermann R, (1984) Process fault detection based on modelling and estimation methods: a survey. Automatica, 20:387–404

Isermann R, (1994) Integration of fault-detection and diagnosis methods. Proc. of IFAC Symposium on Fault Detection, Supervision and Safety for Technical Processes (SAFEPROCESS), Espoo, Finland

Isermann R, (1997) Supervision, fault-detection and fault-diagnosis methods – an introduction. Control Engineering Practice 5(5):639–652

Isermann R, Ballé P, (1997) Trends in the application of model-based fault detection and diagnosis of technical processes. Control Engineering Practice, 5(5):709–719

Omdahl T, (editor) (1988) Reliability, availability and maintainability (RAM) dictionary. ASQC Quality Press, Milwaukee

Poulizeus A, Stravlakakis G, (1994) Real time fault monitoring of industrial processes. Kluwer Academic Publishers, Dordrecht

Willsky AS, (1976) A survey of design methods for failure detection systems. Automatica 12:601–611

Yung SK, Clarke DW, (1989) Sensor validation. Measurement and Control 22:132–150

B.1.6.2 Fault Detection

Alessandri A, Parisini T, (1997) Model-based fault diagnosis using nonlinear estimators: a neural approach. Proc. of the American Control Conference, 2:903–907

Borairi M, Wang H, (1998) Actuator and sensor fault diagnosis of nonlinear dynamic systems via genetic neural networks and adaptive parameter estimation techniques. Proc. of the IEEE International Conference on Control Applications 1:278–282

Chow WY, Willsky AS, (1984) Analytical redundancy and the design of robust failure detection systems. IEEE Transactions on Automatic Control 29:603–614

Demetriou MA, and Polycarpou MM, (1998) Incipient fault diagnosis of dynamical systems using online approximators. IEEE Transactions on Automatic Control 43(11):1612–1617

Emami-Naeini A, (1986) Robust detection, isolation and accommodation for sensor failures. NASA Contractor Report, CR-174825

Gertler J, (1991) Analytical redundancy methods in fault detection and isolation. Proc. of IFAC Symposium on Fault Detection, Supervision and Safety for Technical Processes (SAFEPROCESS), Baden-Baden

Gertler J, (1997) Fault detection and isolation using parity relations. Control Engineering Practice 5(5):653–661

Gertler J, Singer D, (1985) Augmented models for statistical fault isolation in complex dynamic systems. Proc. of American Control Conference (ACC), 1:317–322, Boston, MA, USA

Hoefling T, Isermann R, (1996) Fault detection based on adaptive parity equations and single-parameter tracking. Control Engineering Practice 4(10):1361–1369

Hoefling T, Pfeufer T, (1994) Detection of additive and multiplicative faults – parity space vs. parameter estimation. Proc. of IFAC Symposium on Fault Detection, Supervision and Safety for Technical Processes (SAFEPROCESS), Espoo, Finland

Lou XC, Willsky AS, Verghese GL, (1986) Optimally robust redundancy relations for failure detection in uncertain systems. Automatica 22:333–344

Maki Y, Loparo KA, (1997) A neural-network approach to fault detection and diagnosis in industrial processes. IEEE Transactions on Control Systems Technology 6(5):529–541

Marcu T, Mirea L, (1997) Robust detection and isolation of process faults using neural networks. IEEE Control Systems Magazine 17(5):72–79

Patton RJ, Chen J, (1991) A review of parity space approaches to fault diagnosis. Proc. of IFAC Symposium on Fault Detection, Supervision and Safety for Technical Processes (SAFEPROCESS), Baden-Baden

Patton R, Chen J, (1997) Observer-based fault detection and isolation: robustness and applications. Control Engineering Practice 5(5):671–682

Polycarpou MM, Vemuri AT, (1995) Learning methodology for failure detection and accommodation. IEEE Control Systems Magazine, 6:16–24

Vemuri AT, Polycarpou MM, Diakourtis SA, (1998) Neural network based fault detection in robotic manipulators. IEEE Transactions on Robotics and Automation 14(2):342–348

B.1.6.3 Fault Diagnosis

Ballé P, (1998) Fuzzy model-based symptom generation and fault diagnosis for nonlinear processes. Proc. of the IEEE World Congress on Computational Intelligence 2:945–950

Borairi M, Wang H, (1998) Actuator and sensor fault diagnosis of nonlinear dynamic systems via genetic neural networks and adaptive parameter estimation techniques. Proc. of the IEEE International Conference on Control Applications 1:278–282

Dexter AL, and Benouarets M, (1997) Model-Based fault diagnosis using fuzzy matching. IEEE Transactions on Systems, Man, and Cybernetics, Part A 27(5):673–682

Freyermuth B, (1991) Knowledge-based incipient fault diagnosis of industrial robots. Proc. of IFAC symposium on fault detection, supervision and safety for technical processes (SAFEPROCESS), Baden-Baden

Leonhardt S, Ayoubi M, (1997) Methods of fault diagnosis. Control Engineering Practice 5(5):683–692

Isermann R, (1994) Integration of fault-detection and diagnosis methods. Proc. of IFAC Symposium on Fault Detection, Supervision and Safety for Technical Processes (SAFEPROCESS), Espoo, Finland

Isermann R, (1997) Supervision, fault-detection and fault-diagnosis methods – an introduction. Control Engineering Practice 5(5):639–652

Isermann R, Ulieru E, (1993) Integrated fault detection and diagnosis. Proc. of IEEE/SMC Conference on Systems Engineering in the Service of Humans:743–748. Le Toquet, France

Isermann R, Ulieru E, (1994) On fuzzy logic application for automatic control, supervision and fault diagnosis. Proc. of Third European Conference on Fuzzy and Intelligent Technologies EUFIT, pp.738–753

Milne R, (1987) Strategies for diagnosis. IEEE Transactions on Systems, Man and Cybernetics 17(3):333–339

Polycarpou MM, Vemuri AT, (1995) Learning methodology for failure detection and accommodation. IEEE Control Systems Magazine 6:16–24

Rasmussen J, (1993) Diagnostic reasoning in action. IEEE Trans. on Systems, Man and Cybernetics 23(4):981–992

Struss P, Malik A, Sachenbacher M, (1996) Qualitative modeling is the key to automated diagnosis. Proc. of 13th IFAC World Congress, San Francisco, CA, USA

Torasso P, Console L, (1989) Diagnostic problem solving. North Oxford Academic, UK

Ulieru M, Mrsic-Flogel J, (1994) Integrated neural and fuzzy paradigms for efficient data analysis in diagnosis. Proc. of IFAC Symposium on Fault Detection, Supervision and Safety for Technical Processes (SAFEPROCESS), Espoo, Finland

Zhang J, (2002) Improved on-line process fault diagnosis using stacked neural networks, Proc. of the 2002 IEEE International Conference on Control Applications, Glasgow, UK:689–694

B.1.6.4 Dedicated Conferences

IFAC (1991–2006) Symposium on fault detection, supervision and safety of technical processes (SAFEPROCESS). Pergamon, London

IFAC-Workshop (1986–2001) On-line fault detection and supervision in the chemical process industries. Pergamon, London, 1986–2001

B.2 Applicative Contributions

Ahmad AL, Azid IA, Yusof AR, Seetharamu KN, (2004) Emission control in palm oil mills using artificial neural network and genetic algorithm. Computers and Chemical Engineering 28:2709–2715

Andò B, Cammarata G, Fichera A, Graziani S, Pitrone N, (1999) A procedure for the optimization of air quality monitoring networks. IEEE Transactions on

Systems, Man, and Cybernetics-Part C: Applications and Reviews 29(1):157–163
Assis AJ, Filho RM, (2000) Soft sensors development for on-line bioreactor state estimation. Computers and Chemical Engineering 24:1099–1103
Buceti G, Fortuna L, Rizzo A, Xibilia MG, (2001b) An automatic validation system for interferometry density measurements in the ENEA-FTU tokamak based on soft-computing. 8th International Conference on Accelerator and Large Experimental Physics Control Systems, ICALEPCS01, San Jose, California:343–345
Buceti G, Fortuna L, Rizzo A, Xibilia MG, (2002) Automatic validation of the five-channel interferometer in ENEA-FTU based on soft-computing techniques. Fusion Engineering and Design 60:381–387
Buceti G, Gallo A, Rizzo A, Xibilia MG, (2001) A fuzzy sensor validation system for plasma density measures in tokamak machines based on neural models. SOCO/ISFI 2001, Fourth International ICSC Symposium on Soft Computing and Intelligent Systems for Industry, Paisley, Scotland, UK, June 26–29, 2001
Bucolo M, Fortuna L, Nelke M, Rizzo A, Sciacca T, (2002) Prediction models for the corrosion phenomena in pulp & paper plant. Control Engineering Practice 10:227–237
Caponetto R, Fortuna L, Rizzo A, (2003) Nonlinear modeling of fuel cell systems for vehicles. Nonlinear Phenomena in Complex Dynamics 6(3):746–751
Dorr R, Kratz F, Ragot J, Germain J, (1997) Detection, isolation, and identification of sensor faults in nuclear power plants. IEEE Trans. on Control Systems Technology 5(1):42–60
Dote Y, Ovaska SJ (2001) Industrial applications of soft computing: a review. Proc. of IEEE 89 9:1243–1265
Dufour P, Bhartiya S, Dhurjati PS, Doyle FJ III, (2005) Neural networks-based software sensor: training set design and application to a continuous pulp digester. Control Engineering Practice 13:135–143
Edwards PJ, Murray AF, Papadopoulos G, Wallace AR, Barnard J, Smith G, (1999) The application of neural networks to the papermaking industry. IEEE Transactions on Neural Networks 10(6):1456–1464
Englund C, Verikas A, (2005) A hybrid approach to outlier detection in the offset lithographic printing process. Engineering Applications of Artificial Intelligence 18:759–768
Esposito B, Fortuna L, Rizzo A, (2004) A neural system for radiation discrimination in nuclear fusion applications. Proc. of IEEE International Symposium on Circuits and Systems (ISCAS2004), Vancouver, Canada
Flynn D, Ritchie J, Cregan M, (2005) Data mining techniques applied to power plant performance monitoring. 16th IFAC World Congress, Prague
Fortuna L, Graziani S, Xibilia MG, (2005a) Soft sensors for product quality monitoring in debutanizer distillation columns. Control Engineering Practice 13:499–508
Fortuna L, Graziani S, Xibilia MG, (2005b) Virtual instruments in refineries. IEEE Instrumentation & Measurement Magazine 8:26–34
Fortuna L, Marchese V, Rizzo A, Xibilia MG, (1999) A neural networks based system for post pulse fault detection and data validation in tokamak machines.

Proceedings of IEEE 1999 International Symposium on Circuits and Systems (ISCAS 99) 5:563–566, Orlando, FL, USA

Fortuna L, Rizzo A, Sinatra M, Xibilia MG, (2003) Soft analyzers for a sulfur recovery unit. Control Engineering Practice 11:1491–1500

Govindhasamy JJ, McLoone SF, Irwin GW, French JJ, Doyle RP, (2005) Neural modelling, control and optimisation of an industrial grinding process. Control Engineering Practice 13:1243–1258

Graziani S, Pitrone N, Xibilia MG, Barbalace N, (2004) Improving monitoring of NO_x emissions in refineries. Proc. of IMTC 2004, Como Italy 1:594–597

James S, Legge R, Budman H, (2002) Comparative study of black-box and hybrid estimation methods in fed-batch fermentation. Journal of Process Control 12:113–121

Janes KR, Yang SX, Hacker RR, (2005) Pork farm odour modelling using multiple-component multiple-factor analysis and neural networks. Applied Soft Computing 6:53–61

Komulainen T, Sourander M, Jamsa-Jounela SL, (2004) An online application of dynamic PLS to a dearomatization process. Computers and Chemical Engineering 28:2611–2619

Li K, Peng J, Irwin GW, Piroddi L, Spinelli W, (2005) Estimation of NO_x emissions in thermal power plants using eng-genes neural networks. 16th IFAC World Congress, Prague

Lin B, Recke B, Renaudat P, Knudsen J, Jørgensen SB, (2005) Robust statistics for soft sensor development in cement kiln. 16th IFAC World Congress, Prague

Luo JX, Shao HH, (2006) Developing soft sensors using hybrid soft computing methodology: a neurofuzzy system based on rough set theory and genetic algorithms. Soft Computing 10:54–60

Maione B, Lino P, Rizzo A, (2005) Neural network nonlinear modeling of a common rail injection system for a CNG engine. WSEAS Transactions on Systems 3(5):2282–2287

Matsumura S, Iwahara T, Ogata K, Fujii S, Suzuki M, (1998) Improvement of de-NO_x device control performance using a software sensor. Control Engineering Practice 6:1267–1276

Park S, Han C, (2000) A nonlinear soft sensor based on multivariate smoothing procedure for quality estimation in distillation columns. Computers and Chemical Engineering 24:871–877

Rallo R, Ferre-Giné J, Arena A, Giralt F, (2002) Neural virtual sensor for the inferential prediction of product quality from process variables. Computers and Chemical Engineering 26:17351754

Rizzo A, Xibilia MG, (2002) An innovative intelligent system for sensor validation in tokamak machines. IEEE Transactions on Control System Technology 10(3):421–431

Roverso D, Ruan DA, (2004) Enhancing Cross-Correlation Analysis with Artificial Neural Networks for Nuclear Power Plant Feedwater Flow measurement. Real Time Systems 27:85–96

Ruan D, Roverso D, Fantoni PF, Sanabrias JI, Carrasco JA, Fernandez L, (2003) Integrating cross-correlation techniques and neural networks for feedforward flow measurement. Progress in Nuclear Energy 43(1–4):267–274

Shi Z, Cuimei B, Bin L, Yonghua W, (2005) The advanced process control system for an industrial distillation column. 16th IFAC World Congress, Prague

Su HB, Fan LT, Schlup JR, (1998) Monitoring the process of curing epoxy/graphite fiber composites with a recurrent neural network as a soft sensor. Engineering Applications of Artificial Intelligence 11:293–306

Tham MT, Morris AJ, Montague GA, (1994) Soft-sensing: a solution to the problem of measurement delays. Chemical Engineering Research & Design 67:547–554

van Deventer JSJ, Kam KM, van der Walt TJ, (2004) Dynamic modelling of a carbon-in-leach process with the regression network. Chemical Engineering Science 59: 4575–4589

Wang L, Chessari C, Karpiel E, (2001) Inferential control of product quality attributes-application to food cooking extrusion process. Journal of Process Control 11:621–636

Willis MJ, Montague GA, Di Massimo C, Tham MT, Morris AJ, (1992) Artificial neural networks in process estimation and control. Automatica 28:1181–1187

Xiong Z, Zhang J, (2005) A batch-to-batch iterative optimal control strategy based on recurrent neural network models. Journal of Process Control 15:11–21

Yan W, Shao H, Wang X, (2004) Soft sensing modeling based on support vector machine and Bayesian model selection. Computers and Chemical Engineering 28:1489–1498

Yang SH, Chen BH, Wang XZ, (2000) Neural network based fault diagnosis using unmeasurable inputs. Engineering Applications of Artificial Intelligence 13:345–356

Zahedi G, Elkamel A, Lohi A, Jahanmiri A, Rahimpor MR, (2005) Hybrid artificial neural network – First principle model formulation for the unsteady state simulation and analysis of a packed bed reactor for CO_2 hydrogenation to methanol. Chemical Engineering Journal 115:113–120

Zhang J, (1999) Developing robust non-linear models through bootstrap aggregated neural networks. Neurocomputing 25:93–113

Zhang J, Martin EB, Morris AJ, Kiparissides C, (1997) Inferential estimation of polymer quality using stacked neural networks. Computers and Chemical Engineering 21:S1025–S1030

References

Ahlberg JH, Nilson EN, Walsh JL, (1967) The theory of splines and their applications. Academic Press, New York

Ahmad AL, Azid IA, Yusof AR, Seetharamu KN, (2004) Emission control in palm oil mills using artificial neural network and genetic algorithm. Computers and Chemical Engineering 28:2709–2715

Ahmad Z, Zhang J, (2002) Improving long range prediction for nonlinear process modelling through combining multiple neural networks. Proc. of the 2002 Int. Conference on Control Applications 2: 966 – 971

Albazzaz H, Wang XZ (2006) Historical data analysis based on plots of independent and parallel coordinates and statistical control limits. Journal of Process Control 16:103–114

Alessandri A, Parisini T, (1997) Model-based fault diagnosis using nonlinear estimators: a neural approach. Proc. of the American Control Conference 2:903–907

Andò B, Cammarata G, Fichera A, Graziani S, Pitrone N, (1999) A procedure for the optimization of air quality monitoring networks. IEEE Transactions on Systems, Man, and Cybernetics-Part C: Applications and Reviews 29(1):157–163

Araúzo-Bravo MJ, Cano-Izquierdo JM, Gómez-Sánchez EG, López-Nieto MJ, Dimitriadis YA, López-Coronado J, (2004) Automatization of a penicillin production process with soft sensors and a adaptive controller based on neuro fuzzy systems. Control Engineering Practice 12:1073–1090

Assis AJ, Filho RM, (2000) Soft sensors development for on-line bioreactor state estimation. Computers and Chemical Engineering 24:1099–1103

Baffi G, Martin EB, Morris AJ, (1999) Non-linear projection to latent structures revisited (the neural network PLS algorithm). Computers and Chemical Engineering 23:1293–1307

Ballé P, (1998) Fuzzy model-based symptom generation and fault diagnosis for nonlinear processes. Proc. of the IEEE World Congress on Computational Intelligence 2:945–950

Barlow RE, Proschan F, (1975) Statistical theory of reliability and life testing. Holt, Rinehart and Winston, Inc.

Bezdek JC, (1981) Pattern recognition with fuzzy objective function algorithms. Plenum Press, New York
Billings SA, Jamaluddin HB, Chen S, (1992) Properties of neural networks with applications to modelling non-linear dynamical system. International Journal Control 55:193–224
Billings SA, Voon WSF, (1991) Correlation based model validity tests for nonlinear models. International Journal Control 54:157–194
Bishop CM, (1995) Neural networks for pattern recognition. Clarendon Press, Oxford, UK
Bomberger JD, Seborg DE, (1998) Determination of model order for NARX models directly from input-output data. Journal of Process Control 8:459–468
Borairi M, Wang H, (1998) Actuator and sensor fault diagnosis of nonlinear dynamic systems via genetic neural networks and adaptive parameter estimation techniques. Proc. of the IEEE International Conference on Control Applications 1:278–282
Buceti G, Fortuna L, Rizzo A, Xibilia MG, (2001a) An automatic validation system for interferometry density measurements in the ENEA-FTU tokamak based on soft-computing. 8th International Conference on Accelerator and Large Experimental Physics Control Systems, ICALEPCS01, San Jose, California:343–345
Buceti G, Fortuna L, Rizzo A, Xibilia MG, (2002) Automatic validation of the five-channel interferometer in ENEA-FTU based on soft-computing techniques. Fusion Engineering and Design 60:381–387
Buceti G, Gallo A, Rizzo A, Xibilia MG, (2001b) A fuzzy sensor validation system for plasma density measures in tokamak machines based on neural models. SOCO/ISFI 2001, Fourth International ICSC Symposium on Soft Computing and Intelligent Systems for Industry, Paisley, Scotland, UK, June 26–29, 2001
Bucolo M, Fortuna L, Nelke M, Rizzo A, Sciacca T, (2002) Prediction models for the corrosion phenomena in pulp & paper plant. Control Engineering Practice 10:227–237
Buzio M *et al.*, (1996) Axisymmetric and non-axisymmetric structural effects of disruption-induced electromechanical forces on the JET tokamak. Fusion Technology 27:755–758
Caponetto R, Fortuna L, Rizzo A, (2003) Nonlinear modeling of fuel cell systems for vehicles. Nonlinear Phenomena in Complex Dynamics 6(3):746–751
Chen J, Patton R, (1999) Robust model-based fault diagnosis for dynamic systems. Kluwer, Boston
Chen S, Billings SA, (1989) Representations of non-linear system: the NARMAX model. International Journal of Control 49:1012–1032
Chen S, Billings SA, Grant PM, (1990) Non-linear system identification using neural networks. International Journal of Control 51:1191–1214
Chen S, Cowan CFN, Grant PM, (1991) Orthogonal least squares algorithms for training multi-output radial basis function networks. Proc. of 2nd International Conference on Artificial Neural Networks, Bournemouth, UK 1:336–339
Chiang LH, Perl RJ, Seasholtz MB, (2003) Exploring process data with the use of robust outlier detection algorithm. Journal of Process Control 13:437:449

Chiang LH, Russell EL, Braatz RD, (2001) Fault detection and diagnosis in industrial systems. Advanced Books in Control and Signal Processing Series, Springer-Verlag, London

Chow WY, Willsky AS, (1984) Analytical redundancy and the design of robust failure detection systems. IEEE Transactions on Automatic Control 29:603–614

Combs Clide F Jr, (1999) Electronic instrument handbook, McGraw-Hill Handbooks: 45.1–45.24

Dadhe K, Engell S, (2005) Assessing the predictions of dynamic neural networks. 16th IFAC World Congress, Prague

Demetriou MA, and Polycarpou MM, (1998) Incipient fault diagnosis of dynamical systems using online approximators. IEEE Transactions on Automatic Control 43(11):1612–1617

Dexter AL, and Benouarets M, (1997) Model-based fault diagnosis using fuzzy matching. IEEE Transactions on Systems, Man, and Cybernetics, Part A 27(5):673–682

Dorr R, Kratz F, Ragot J, Germain J, (1997) Detection, isolation, and identification of sensor faults in nuclear power plants. IEEE Trans. on Control Systems Technology 5(1):42–60

Dote Y, Ovaska SJ (2001) Industrial applications of soft computing: a review. Proc. of IEEE 89, 9:1243–1265

Dufour P, Bhartiya S, Dhurjati PS, Doyle FJ III, (2005) Neural networks-based software sensor: training set design and application to a continuous pulp digester. Control Engineering Practice 13:135–143

Edwards PJ, Murray AF, Papadopoulos G, Wallace AR, Barnard J, Smith G, (1999) The application of neural networks to the papermaking industry. IEEE Transactions on Neural Networks 10(6):1456–1464

Emami-Naeini A, (1986) Robust detection, isolation and accommodation for sensor failures. NASA Contractor Report, CR-174825

Englund C, Verikas A, (2005) A hybrid approach to outlier detection in the offset lithographic printing process. Engineering Applications of Artificial Intelligence 18:759–768

Esposito B, Fortuna L, Rizzo A, (2004) A neural system for radiation discrimination in nuclear fusion applications. Proc. of IEEE International Symposium on Circuits and Systems (ISCAS2004), Vancouver, Canada

Eykhoff P, (1974) System identification. John Wiley, London

Feil B, Abonyi J, Szeifert F, (2004) Model order selection of nonlinear input-output models – a clustering based approach. Journal of Process Control 14:593–602

Feng S, Chen J, Tu XY, (2005) Nonlinear system identification with shortage of input-output data. 16th IFAC World Congress, Prague

Flynn D, Ritchie J, Cregan M, (2005) Data mining techniques applied to power plant performance monitoring. 16th IFAC World Congress, Prague

Fortuna L, Graziani S, Xibilia MG, (2005a) Soft sensors for product quality monitoring in debutanizer distillation columns. Control Engineering Practice 13:499–508

Fortuna L, Graziani S, Xibilia MG, (2005b) Virtual instruments in refineries. IEEE Instrumentation & Measurement Magazine 8:26–34

Fortuna L, Marchese V, Rizzo A, Xibilia MG, (1999) A neural networks based system for post pulse fault detection and data validation in tokamak machines. Proceedings of IEEE 1999 International Symposium on Circuits and Systems (ISCAS 99), 5:563–566, Orlando, FL, USA

Fortuna L, Rizzo A, Sinatra M, Xibilia MG, (2003) Soft analyzers for a sulfur recovery unit. Control Engineering Practice 11:1491–1500

Fortuna L, Rizzotto G, Lavorgna M, Nunnari G, Xibilia MG, Caponetto R, (2001) Soft computing: new trends and applications. Advanced Textbooks in Control and Signal Processing, Springer

Foster KR, (1998) Software tools, IEEE Spectrum 53,1: 52–56

Frank PM, (1990) Fault diagnosis in dynamic systems using analytical and knowledge-based redundancy – a survey and some new results. Automatica 26(3):459–474

Freyermuth B, (1991) Knowledge-based incipient fault diagnosis of industrial robots. proc. of ifac symposium on fault detection, supervision and safety for technical processes (SAFEPROCESS), Baden-Baden

Frost RA, (1986) Introduction to knowledge base systems. Collins, London, UK

Gertler J, (1988) Survey of model-based failure detection and isolation in complex plants. IEEE System and Control Magazine, December 1988:3–17

Gertler J, (1991) Analytical redundancy methods in fault detection and isolation. Proc. of IFAC Symposium on Fault Detection, Supervision and Safety for Technical Processes (SAFEPROCESS), Baden-Baden

Gertler J, (1997) Fault detection and isolation using parity relations. Control Engineering Practice 5(5):653–661

Gertler J, (1998) Fault detection and diagnosis in engineering systems. Marcel Dekker, New York

Gertler J, Singer D, (1985) Augmented models for statistical fault isolation in complex dynamic systems. Proc. of American Control Conference (ACC), 1:317–322, Boston, MA, USA

Govindhasamy JJ, McLoone SF, Irwin GW, French JJ, Doyle RP, (2005) Neural modelling, control and optimisation of an industrial grinding process. Control Engineering Practice 13:1243–1258

Graziani S, Pitrone N, Xibilia MG, Barbalace N, (2004) Improving monitoring of NO_x emissions in refineries. Proc. of IMTC 2004, Como Italy 1:594–597

Guidorzi R, (2003) Multivariable System Identification. Bonomia University Press, Bologna

Gupta MM, Sinha NK, (2000) Soft Computing and Intelligent Systems: Theory and Applications. Academic Press Series in Engineering, San Diego CA

Haber R, Unbehahuen H, (1990) Structure identification of nonlinear dynamic systems – a survey on input/output approaches. Automatica 26(4):651–677

Hashem S, (1997) Optimal linear combinations of neural networks. Neural Networks 10(4):599–614

Haykin S, (1999) Neural networks: a comprehensive foundation, 2nd edn. Prentice Hall

He X, Asada H, (1993) a new method for identifying orders of input–output models for nonlinear dynamic systems. Proc. Of American control Conference, San Francisco CA, 3:2520–2522

Henry MP, Clarke DW, (1993) The self validating sensor: rationale, definitions, and examples. Control Engineering Practice 1(4):585–610

Himmelblau DM, (1978) Fault detection and diagnosis in chemical and petrochemical processes. Elsevier, Amsterdam

Hoefling T, Isermann R, (1996) Fault detection based on adaptive parity equations and single-parameter tracking. Control Engineering Practice 4(10):1361–1369

Hoefling T, Pfeufer T, (1994) Detection of additive and multiplicative faults – parity space vs. parameter estimation. Proc. of IFAC Symposium on Fault Detection, Supervision and Safety for Technical Processes (SAFEPROCESS), Espoo, Finland

Hutchinson IH, (1990) Principles of plasma diagnostics. Cambridge University Press

IFAC (1991–2006) Symposium on fault detection, supervision and safety of technical processes (SAFEPROCESS). Pergamon, London

IFAC-Workshop (1986–2001) On-line fault detection and supervision in the chemical process industries. Pergamon, London, 1986–2001

Isermann R, (1984) Process fault detection based on modelling and estimation methods: a survey. Automatica 20:387–404

Isermann R, (1994) Integration of fault-detection and diagnosis methods. Proc. of IFAC Symposium on Fault Detection, Supervision and Safety for Technical Processes (SAFEPROCESS), Espoo, Finland

Isermann R, (1997) Supervision, fault-detection and fault-diagnosis methods – an introduction. Control Engineering Practice 5(5):639–652

Isermann R, (2006a) Fault-diagnosis systems, an introduction from fault detection to fault tolerance, Springer-Verlag, Berlin, Heidelberg

Isermann R, (2006b) Fault diagnosis of technical processes – applications. Springer-Verlag, Berlin, Heidelberg

Isermann R, Ballé P, (1997) Trends in the application of model-based fault detection and diagnosis of technical processes. Control Engineering Practice 5(5):709–719

Isermann R, Ulieru E, (1993) Integrated fault detection and diagnosis. Proc. of IEEE/SMC Conference on Systems Engineering in the Service of Humans:743–748. Le Toquet, France

Isermann R, Ulieru E, (1994) On fuzzy logic application for automatic control, supervision and fault diagnosis. Proc. of Third European Conference on Fuzzy and Intelligent Technologies EUFIT, pp.738–753

Janes KR, Yang SX, Hacker RR, (2005) Pork farm odour modelling using multiple-component multiple-factor analysis and neural networks. Applied Soft Computing 6:53–61

James S, Legge R, Budman H, (2002) Comparative study of black-box and hybrid estimation methods in fed-batch fermentation. Journal of Process Control 12:113–121

Jang JR, Sun C, Mizutani E, (1997) Neuro fuzzy and soft computing: a computational approach to learning and machine intelligence. Prentice Hall.

JET Joint Undertaking (1998) 1998 JET operation instructions. JET Joint Undertaking, Internal Documentation, Culham, UK

Juditsky A, Hjalmarsson H, Benveniste A, Delyon B, Ljung L, Sjoberg J, Zhang Q, (1995) Nonlinear black-box models in system identification: mathematical foundations. Automatica 31:1725–1750

Kahrs O, Brendel M, Marquardt W, (2005) Incremental identification of NARX models by sparse grid approximation. 16th IFAC World Congress, Prague

Komulainen T, Sourander M, Jamsa-Jounela SL, (2004) An online application of dynamic PLS to a dearomatization process. Computers and Chemical Engineering 28:2611–2619

Kreyszig E, (1999) Advanced engineering mathematics: international edition. John Wiley & Sons Ltd

Lang ZQ, Futterer M, Billings SA, (2005) The identification of a class of nonlinear systems using a correlation analysis approach. 16th IFAC World Congress, Prague

Lee JM, Yoo C, Lee IB, (2004) Statistical process monitoring with independent component analysis. Journal of Process Control 14:467–485

Leonhardt S, Ayoubi M, (1997) Methods of fault diagnosis. Control Engineering Practice 5(5):683–692

Levine W.S. (editor) (1996) The Control Handbook: 562–565. CRC Press – IEEE Press, Boca Raton, FL, USA.

Li K, Peng J, Irwin GW, Piroddi L, Spinelli W, (2005) Estimation of NO_x emissions in thermal power plants using eng-genes neural networks. 16th IFAC World Congress, Prague

Lin B, Recke B, Renaudat P, Knudsen J, Jørgensen SB, (2005) Robust statistics for soft sensor development in cement kiln. 16th IFAC World Congress, Prague

Lind I, (2005) Nonlinear structure identification with linear least squares and ANOVA. 16th IFAC World Congress, Prague

Lind I, Ljung L, (2005) Regressor selection with the analysis of variance method. Automatica 41:693–700

Lipmann RP, (1987) An introduction to computing with neural nets. IEEE ASSP Magazine, April:4–22

Liu J, (2005) On line soft sensor for polyethylene process with multiple production grades. 16th IFAC World Congress, Prague

Ljung L, (1999) System identification: theory for the user, 2nd edn. T. Kailath, Ed. Englewood Cliffs, Prentice Hall

Looney CG, (1997) Pattern recognition using neural networks – theory and algorithms for engineers and scientists. Oxford University Press, Oxford, UK

Lopes VV, Menezes JC, (2005) Inferential sensor design in the presence of missing data: a case study. Chemometrics and Intelligent Laboratory System 78:1–10

Lou XC, Willsky AS, Verghese GL, (1986) Optimally robust redundancy relations for failure detection in uncertain systems. Automatica 22:333–344

Luo JX, Shao HH, (2006) Developing soft sensors using hybrid soft computing methodology: a neurofuzzy system based on rough set theory and genetic algorithms. Soft Computing 10:54–60

Maione B, Lino P, Rizzo A, (2005) Neural network nonlinear modeling of a common rail injection system for a CNG engine. WSEAS Transactions on Systems 3(5):2282–2287

Maki Y, Loparo KA, (1997) A Neural-network approach to fault detection and diagnosis in industrial processes. IEEE Transactions on Control Systems Technology 6(5):529–541

Marchese *et al.*, (1997) Enhancement of JET machine instrumentation and coil protection systems. Fusion Technology, Elsevier Science:747–750

Marcu T, Mirea L, (1997) Robust detection and isolation of process faults using neural networks. IEEE Control Systems Magazine, 17(5):72–79

Masson MH, Canu S, Grandvalet Y, Lynggaard-Jensen A, (1999) Software sensor design based on empirical data. Ecological Modelling 120:131–139

Matsumura S, Iwahara T, Ogata K, Fujii S, Suzuki M, (1998) Improvement of de-NO_x device control performance using a software sensor. Control Engineering Practice 6:1267–1276

Mendes E, Billings SA, (2001) An alternative solution to the model structure selection problem. IEEE Transactions on System Man and Cybernetics Part A-System and Humans 31:597–608

Milne R, (1987) Strategies for diagnosis. IEEE Transactions on Systems, Man and Cybernetics 17(3):333–339

Nagai EY, Arruda LVR, (2005) Soft sensor based on fuzzy model identification. 16th IFAC World Congress, Prague

Narendra KS, Parthasarathy K, (1990) Identification and control of dynamical systems using neural networks. IEEE Transactions on Neural Networks 1(1):4–27

Nelson PRC, Taylor PA, MacGregor JF, (1996) Missing data methods in PCA and PLS: score calculations with incomplete observations. Chemometrics and Intelligent Laboratory Systems 35(1):45–65

NIST/SEMATECH (2005) e-Handbook of Statistical Methods, http://www.itl.nist.gov/div898/handbook/

Nørgaard M, Ravn O, Poulsen NK, Hansen LK, (2000) Neural networks for modelling and control of dynamic system. Springer, London

Omdahl T, (editor) (1988) Reliability, availability and maintainability (RAM) dictionary. ASQC Quality Press, Milwaukee

Omidvar O, Elliot DL, (1997) Neural systems for control. Academic Press, San Diego, CA, USA.

Oppenheim AV, Schafer RW, (1989) Discrete-time signal processing. Prentice-Hall

Papadopoulos G, Edwards PJ, Murray AF, (2001) Confidence estimation methods for neural networks: a practical comparison. IEEE Transactions on Neural Networks 12(6):1456–1464

Park S, Han C, (2000) A nonlinear soft sensor based on multivariate smoothing procedure for quality estimation in distillation columns. Computers and Chemical Engineering 24:871–877

Patton RJ, Chen J, (1991) A review of parity space approaches to fault diagnosis. Proc. of IFAC Symposium on Fault Detection, Supervision and Safety for Technical Processes (SAFEPROCESS), Baden-Baden

Patton R, Chen J, (1997) Observer-based fault detection and isolation: robustness and applications. Control Engineering Practice 5(5):671–682

Patton R, Frank P, Clark R, (1989) Fault diagnosis in dynamic systems. Prentice Hall, Hertfordshire, UK

Patton R, Frank P, Clark P, (eds.) (2000) Issues of fault diagnosis for dynamic systems. Springer, New York

Pau LF, (1981) Failure diagnosis and performance monitoring. Marcel Dekker, New York

Pearl J, (1988) Probabilistic reasoning in intelligent systems: networks of plausible inference. Morgan Kaufmann, San Mateo, California

Pearson RK, (2002) Outliers in process modeling and identification. IEEE Transactions on Control Systems Technology 10(1):55–63

Polycarpou MM, Vemuri AT, (1995) Learning methodology for failure detection and accommodation. IEEE Control Systems Magazine 6:16–24

Ponton JW, Klemes J, (1993) Alternatives to Neural Networks for Inferential Measurement. Computers and Chemical Engineering 17(10):991–1000

Poulizeus A, Stravlakakis G, (1994) Real time fault monitoring of industrial processes. Kluwer Academic Publishers, Dordrecht

Quek CI, Balasubramanian R, Rangaiah GP, (2000) Consider using 'soft analysers' to improve SRU control. Hydrocarbon Processing 79(1):101–106

Rallo R, Ferre-Giné J, Arena A, Giralt F, (2002) Neural virtual sensor for the inferential prediction of product quality from process variables. Computers and Chemical Engineering 26:1735–1754

Rasmussen J, (1993) Diagnostic reasoning in action. IEEE Trans. on Systems, Man and Cybernetics 23(4):981–992

Rizzo A, Xibilia MG, (2002) An innovative intelligent system for sensor validation in tokamak machines. IEEE Transactions on Control System Technology 10(3):421–431

Roverso D, Ruan DA, (2004) Enhancing Cross-Correlation Analysis with Artificial Neural Networks for Nuclear Power Plant Feedwater Flow measurement. Real Time Systems 27:85–96

Ruan D, Roverso D, Fantoni PF, Sanabrias JI, Carrasco JA, Fernandez L, (2003) Integrating cross-correlation techniques and neural networks for feedforward flow measurement. Progress in Nuclear Energy 43(1–4):267–274

Simani S, Fantuzzi C, Patton RJ, (2002) Model-based fault diagnosis in dynamic systems using identification techniques. Advances in Industrial Control Series, Springer, London

Shi Z, Cuimei B, Bin L, Yonghua W, (2005) The advanced process control system for an industrial distillation column. 16th IFAC World Congress, Prague

Sjoberg J, Hjalmarsson H, Ljung L, (1994) Neural networks in system identification. 10th IFAC System Identification-Symposium, Cophenagen

Sjoberg J, Zhang Q, Ljung L, Benveniste A, Delyon B, Glorennec PY, Hjalmarsson H, Juditsky A, (1995) Nonlinear black-box modeling in system identification: a unified overview. Automatica 31:1691–1724

Spinelli W, Piroddi L, Lovera M, (2005) On the role of pre–filtering in nonlinear system identification. 16th IFAC World Congress, Prague

Sridhar DV, Bartlett EB, Seagrave RC, (1999) An information theoretic approach for combining neural network process models. Neural Networks 12:915–926

Stern AC, (1982) History of air pollution legislation in the united states. Journal of the Air Pollution Control Association 32:44–61

Stoica P, Södersröm T, (1989) System identification. Englewood Cliff, NJ, Prentice Hall

Struss P, Malik A, Sachenbacher M, (1996) Qualitative modeling is the key to automated diagnosis. Proc. of 13th IFAC World Congress, San Francisco, CA, USA

Su HB, Fan LT, Schlup JR, (1998) Monitoring the process of curing epoxy/graphite fiber composites with a recurrent neural network as a soft sensor. Engineering Applications of Artificial Intelligence 11:293–306

Takagi T, Sugeno M, (1985) Fuzzy identification of systems and its applications to modelling and control. IEEE Transactions on Systems, Man, and Cybernetics 5(3):116–132.

Tham MT, Morris AJ, Montague GA, (1994) Soft-sensing: a solution to the problem of measurement delays. Chemical Engineering Research & Design 67:547–554

Thornhill NF, Shoukat Choudhury MAA, Shah SL, (2004) The impact of compression on data-driven process analyses. Journal of Process Control 14:389–398

Torasso P, Console L, (1989) Diagnostic problem solving. North Oxford Academic, UK

Tou JT, Gonzales RC, (1984) Pattern recognition principles. Addison-Wesley Publishing, Reading, Massachusetts, USA

Ulieru M, Mrsic-Flogel J, (1994) Integrated neural and fuzzy paradigms for efficient data analysis in diagnosis. Proc. of IFAC Symposium on Fault Detection, Supervision and Safety for Technical Processes (SAFEPROCESS), Espoo, Finland

van Deventer JSJ, Kam KM, van der Walt TJ, (2004) Dynamic modelling of a carbon-in-leach process with the regression network. Chemical Engineering Science 59: 4575–4589

Vemuri AT, Polycarpou MM, Diakourtis SA, (1998) Neural network based fault detection in robotic manipulators. IEEE Transactions on Robotics and Automation 14(2):342–348

Wang L, Chessari C, Karpiel E, (2001) Inferential control of product quality attributes-application to food cooking extrusion process. Journal of Process Control 11:621–636

Warne K, Prasad G, Rezvani S, Maguire L, (2004) Statistical and computational intelligence techniques for inferential model development: a comparative evaluation and a novel proposition for fusion. Engineering Applications of Artificial Intelligence 17:871–930

Wesson J, (1987) Tokamaks. Clarendon Press, Oxford, UK

Willis MJ, Montague GA, Di Massimo C, Tham MT, Morris AJ, (1992) Artificial neural networks in process estimation and control. Automatica 28:1181–1187

Willsky AS, (1976) A survey of design methods for failure detection systems. Automatica 12:601–611

Wolpert DH, (1992) Stacked generalization. Neural Networks 5: 241–259

Xiong Z, Zhang J, (2005) A batch-to-batch iterative optimal control strategy based on recurrent neural network models. Journal of Process Control 15:11–21

Yan W, Shao H, Wang X, (2004) Soft sensing modeling based on support vector machine and Bayesian model selection. Computers and Chemical Engineering 28:1489–1498

Yang SH, Chen BH, Wang XZ, (2000) Neural network based fault diagnosis using unmeasurable inputs. Engineering Applications of Artificial Intelligence 13:345–356

Yung SK, Clarke DW, (1989) Sensor validation. Measurement and Control 22:132–150

Zahedi G, Elkamel A, Lohi A, Jahanmiri A, Rahimpor MR, (2005) Hybrid artificial neural network – First principle model formulation for the unsteady state simulation and analysis of a packed bed reactor for CO_2 hydrogenation to methanol. Chemical Engineering Journal 115:113–120

Zhang H, Zouaoui Z, Lennox B, (2005) A comparative study of soft-sensing methods for fed-batch fermentation processes. 16th IFAC World Congress, Prague

Zhang J, (1999) Developing robust non-linear models through bootstrap aggregated neural networks. Neurocomputing 25:93–113

Zhang J, (2002) Improved on-line process fault diagnosis using stacked neural networks, Proc. of the 2002 IEEE International Conference on Control Applications, Glasgow, UK:689–694

Zhang J, Martin EB, Morris AJ, Kiparissides C, (1997) Inferential estimation of polymer quality using stacked neural networks. Computers and Chemical Engineering 21:S1025–S1030

Index

Other titles published in this Series (continued):

Modelling and Analysis of Hybrid Supervisory Systems
Emilia Villani, Paulo E. Miyagi and Robert Valette
Publication due November 2006

Process Control Performance Assessment
Andrzej Ordys, Damien Uduehi and Michael A. Johnson (Eds.)
Publication due December 2006

Model-based Process Supervision
Belkacem Ould Bouamama and Arun K. Samantaray
Publication due February 2007

Magnetic Control of Tokamak Plasmas
Marco Ariola and Alfredo Pironti
Publication due May 2007

Continuous-time Model Identification from Sampled Data
Hugues Garnier and Liuping Wang (Eds.)
Publication due May 2007

Process Control
Jie Bao, and Peter L. Lee
Publication due June 2007

Distributed Embedded Control Systems
Matjaž Colnarič, Domen Verber and Wolfgang A. Halang
Publication due October 2007

Optimal Control of Wind Energy Systems
Iulian Munteanu, Antoneta Iuliana Bratcu, Nicolas-Antonio Cutululis and Emil Ceanga
Publication due November 2007

Analysis and Control Techniques for Distribution Shaping in Stochastic Processes
Michael G. Forbes, J. Fraser Forbes, Martin Guay and Thomas J. Harris
Publication due January 2008

Printed in the United States
65245LVS00002B/169-183